E.coli

Other books in Clay's Library of Health and the Environment:

Air Quality Assessment and Management – *Owen Harrop* 0415234107 (HB), 0415234115 (PB)

Environmental Health and Housing – *Jill Stewart* 041525129X

Also available from Spon Press:

Air Pollution – *Jeremy Colls* 0419206507

Clay's Handbook of Environmental Health 18th Edition – *edited by W. H. Bassett* 0419229604

Decision-making in Environmental Health – *edited by C. Corvalán, D. Briggs and G. Zielhuis* 0419259406 HB, 0419259503 PB

Environmental Health Procedures 5th Edition – *W. H. Bassett* 0419229701

Groundwater Quality Monitoring – *S. Foster, P. Chilton and R. Helmer* 0419258809 HB, 0419258906 PB

Legal Competence in Environmental Health – *Terence Moran* 0419230009

Monitoring Bathing Waters – *edited by Jamie Bartram and Gareth Rees* 0419243704 HB, 0419243801 PB

Statistical Methods in Environmental Health – *J. Pearson and A. Turton* 0412484501

Toxic Cyanobacteria in Water – *edited by Ingrid Chorus and Jamie Bartram* 0419239308

Upgrading Water Treatment Plants – *Glen Wagner and Renato Pinheiro* 0419260404 HB, 0419260501 PB

Water Pollution Control – *Richard Helmer and Ivanildo Hespanhol* 0419229108

Water Quality Assessments – *Deborah Chapman* 0419215905 HB, 0419216006 PB

Water Quality Monitoring – *Jamie Bartram and Richard Balance* 0419223207 HB, 0419217304 PB

Urban Traffic Pollution – *D. Schwela and O. Zali* 0419237208

E.coli

Environmental health issues of VTEC O157

Sharon Parry and Stephen Palmer

SPON PRESS
Taylor & Francis Group

London and New York

First published 2002 by Spon Press
11 New Fetter Lane, London EC4P 4EE

Silmultaneously published in the USA and Canada
by Spon Press
29 West 35th Street, New York, NY 10001

Spon Press is an imprint of the Taylor & Francis Group

Typeset in Palatino by Keystroke,
Jacaranda Lodge, Wolverhampton
Printed and bound in Great Britain by Biddles Ltd,
Guildford and King's Lynn

British Library Cataloguing in Publication Data
A catalogue record for this book is available
from the British Library

Library of Congress Cataloging in Publication Data
Parry, Sharon.
E.coli : environmental health issues of VTEC O157 /
Sharon Parry and Stephen Palmer.
p. cm. — (Clay's library of health and the environment)
Includes bibliographical references and index.
1. Escherichia coli O157:H7. 2. Escherichia coli infections.
3. Environmental health. I. Palmer, Stephen R.
II. Title. III. Series.

RA644.E83 P374 2002
616′.01442—dc21 2001049356

ISBN 0–415–23595–2

Clay's Library of Health and the Environment

An increasing breadth and depth of knowledge is required to tackle the health threats of the environment in the 21st century, and to accommodate the increasing sophistication and globalisation of policies and practices.

Clay's Library of Health and the Environment provides a focus for the publication of leading-edge knowledge in this field, tackling broad and detailed issues. The flagship publication *Clay's Handbook of Environmental Health,* now in its 18th edition, continues to serve environmental health officers and other professionals in over thirty countries.

Contents

Acknowledgements x
Abbreviations xi

1 Overview 1

E.coli *which cause disease 2*
The emergence of VTEC O157 4
How VTEC causes illness in humans 5
Incubation period 8
Infectious dose 9
Practical implications of virulence properties 9
The disease 10
Factors influencing the emergence of VTEC 13
Summary 16
Key messages 16

2 How is VTEC diagnosed? 17

Examination of faecal specimens 17
Isolating E.coli *O157 17*
Identifying toxin production 19
Strain discrimination 21
Serology 22
Period of excretion 23
Future developments 25
Summary 25
Key messages 26

3 How common is VTEC infection? 27

Population based surveillance 27
HUS surveillance 30
Incidence in the UK 31

Detecting outbreaks 36
Conclusions 36
Special studies 37
Summary 40
Key message 40

4 How does VTEC spread to humans? Evidence from sporadic cases 41

Introduction 41
Finding the sources of sporadic infections 41
Analysis and interpretation of case control studies 43
Case control studies of sporadic VTEC O157 infection 47
Microbiological investigations of sporadic cases 49
Investigating sporadic cases of VTEC O157 50
Key messages 56

5 How does VTEC spread to humans? Evidence from outbreaks 58

Food and drinking water outbreaks – beef products 58
Food and drinking water outbreaks – milk and dairy products 60
Food and drinking water outbreaks – other foods 62
Food and drinking water outbreaks – food outlets 63
Food and drinking water outbreaks – drinking water 68
Animal outbreaks 69
Person-to-person spread outbreaks 70
Outbreaks from recreational exposures 70
A practical approach to the investigation of an outbreak 71
Why are VTEC O157 outbreaks different from other food poisoning outbreaks? 74
Summary 75
Key messages 75

6 Animal and environmental reservoirs of VTEC 76

Farm animals 76
Food and water 79
Summary 82
Key messages 82

7 Control of VTEC O157 83

Introduction 83
Background on each of the organisations 83

Control point 1: On-farm control 85
Control point 2: Contamination of raw meat 89
Control point 3: Contamination of milk 93
Control point 4: Contamination of crops and produce (fruit and vegetables) by animal manure 97
Control point 5: Contamination of crops and produce (fruit and vegetables) by human faeces 101
Control point 6: Contamination of drinking water by animal manure 104
Control point 7: Contamination of recreational water by animal manure 110
Control point 8: Contamination of drinking water by human faeces 114
Control point 9: Contamination of recreational water by human faeces 120
Control point 10: Zoonotic spread 125
Control point 11: Consumption of raw and undercooked meat and other food products 129
Control point 12: Consumption of ready-to-eat foods cross-contaminated by VTEC O157 138
Control point 13: Person-to-person transmission via infected food handlers or faecal–oral transmission 145

Glossary 152
Useful contact details 158
Bibliography 163
Index 171

Acknowledgements

The authors wish to express their sincere thanks to colleagues who were consulted in the preparation of this text and to Mrs Ruth Coomber for her invaluable administrative assistance.

Abbreviations

CDC	Centers for Disease Control and Prevention (USA)
E.coli	*Escherichia coli*
EHO	Environmental Health Officer
FSA	Food Standards Agency
HUS	Haemolytic Uraemic Syndrome
LCDC	Laboratory Center for Disease Control (Canada)
LT	Heat labile toxin (produced by some *E.coli*)
MAFF	Ministry of Agriculture, Fisheries and Food
MHS	Meat Hygiene Service
PFGE	Pulsed field gel electrophoresis
PHLS	Public Health Laboratory Service
PHLS – CDSC	Public Health Laboratory Service Communicable Disease Surveillance Centre
PHLS – CDSC (Welsh Unit)	Public Health Laboratory Service Communicable Disease Surveillance Centre (Welsh Unit)
PHLS – LEP	Public Health Laboratory Service Laboratory of Enteric Pathogens
RFLP	Restriction fragment length polymorphism
ST	Heat stable toxin (produced by some *E.coli*)
VT	Vero cytotoxin
VTEC	Vero cytotoxin producing *Escherichia coli*
VTEC O157	Vero cytotoxin producing *Escherichia coli* O157

Chapter 1

Overview

Escherichia coli (*E.coli*) bacteria are common commensals of the gastro-intestinal tract of humans and animals. There are many different strains of *E.coli* and only a few cause disease. This book is about those strains of *E.coli* which produce toxins known as vero cytotoxins which damage the intestinal tract, and which in some people cause serious kidney failure.

Several serotypes of *E.coli* can produce these toxins and they are collectively known as VTEC. However, in the UK the principal cause of VTEC disease is *E.coli* O157. There is no specific treatment for this: antibiotic treatments have been used but there is no indication from several studies that they reduce the length of illness, therefore prevention is particularly important. In comparison with other pathogens which cause diarrhoea in the UK (Figure 1.2) the incidence of VTEC O157 is very low. On average, an Environmental Health Officer (EHO) is likely to investigate less than one sporadic case per year and is very unlikely to ever have to investigate an outbreak. He or she is therefore unlikely to be an expert on the

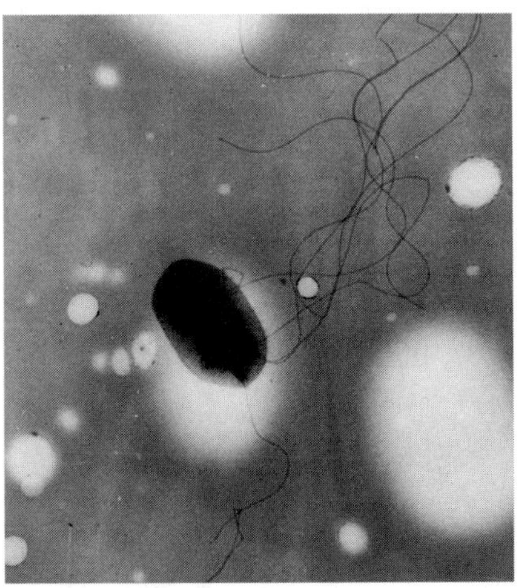

Figure 1.1 *E.coli* bacteria

investigation and control of the disease. However, when cases do occur, the EHO plays a key role in investigating and preventing further spread. The following text provides easily accessible, practical information.

E.colis which cause disease

Typing of *E.coli* is based on the identification of lipopolysaccharide capsular antigens, the somatic 'O' antigens and the flagellar H antigens. VTEC O157 possesses 'O' antigens 157 an 'H' antigen 7. Together they constitute the serotype O157:H7. *E.coli* which caused diarrhoea can also be categorised on the basis of the mechanisms by which they cause the diarrhoea (Table 1.1).

Enteropathogenic (EPEC): The first of these was recognised in 1940. This infection has been associated with outbreaks of diarrhoea in young children. Illness is caused by the bacteria adhering to the intestinal wall.

Table 1.1 Categories of diarrhoeagenic *E.coli*

Category	*Virulence mechanism*	*Examples of predominant O serogroups*	*Type of diarrhoea*
Enteropathogenic (EPEC)	3 subclasses of attachment mechanisms[1]	111, 126, 142, 114 [11]	mainly acute diarrhoea in children under 6 months[11]
Enterotoxigenic (ETEC)	heat stable (ST) or heat labile (LT) toxin[2 3]	1, 6, 8, 11, 114 [10]	acute watery [3]
Enteroinvasive (EIEC)	cell invasion[4 5]	11, 28, 29, 112 [10]	acute dysenteric[12]
Enterohaemorrhagic (EHEC)	attachment, attachment and effacement and vero cytotoxin production[6 7 8]	157, 26, 111, 113 [10]	bloody diarrhoea (HUS)[13 14]
Enteroaggregative (EAggEC)	aggregative adherence	51, 78, 111 [10]	persistent [15]
Diffusely adherent (DAEC)	fimbrial adhesion[9]	75, 126 [10]	? persistent [16]

[1]Donnenberg and Kaper, 1992; [2]Levine *et al.*, 1983; [3]Levine, 1987; [4]Harris *et al.*, 1982; [5]Hale *et al.*, 1983; [6]Karch *et al.*, 1987; [7]Tzipori *et al.*, 1987; [8]O'Brien *et al.*, 1983; [9]Bilge *et al.*, 1989; [10]Guerrant and Thielman, 1995; [11]Levine and Edelman, 1984; [12]Marier *et al.*, 1973; [13]Riley *et al.*, 1983; [14]Karmali *et al.*, 1985;[15]Bhan *et al.*, 1989;[16]Baqui *et al.*, 1992.

Enterotoxigenic (ETEC): This category came to prominence in the late 1960s and was associated with infant diarrhoea in developing countries as well as traveller's diarrhoea. The acute watery diarrhoea is produced by heat stable or heat label toxins.

Enteroinvasive (EIEC): Enteroinvasive *E.coli* were first described in 1971 and are very closely related to Shigella bacteria causing dysentery type illness in humans by the invasion of the gut epithelial cells.

Enterohaemorrhagic (EHEC): The emergence of VTEC in the early 1980s led to the recognition of enterohaemorrhagic *E.coli* associated with haemorrhagic colitis and haemolytic uraemic syndrome. Other EHEC have been recognised which are all vero cytotoxin producing. Vero cytotoxin is a toxin which destroys vero cells, a particular laboratory cell line, and also affects the cells lining the gastrointestinal tract and the kidneys. Further information on how VTEC cause disease is given later in this chapter.

Enteroaggregative (EaggEC): More recently enteroaggregative *E.coli* have been associated with persistent diarrhoea in people in developing countries. They have been isolated in some patients in the UK. A further group, called diffusely adherent *E.coli*, has been identified and has caused persistent childhood diarrhoea in Bangladesh.

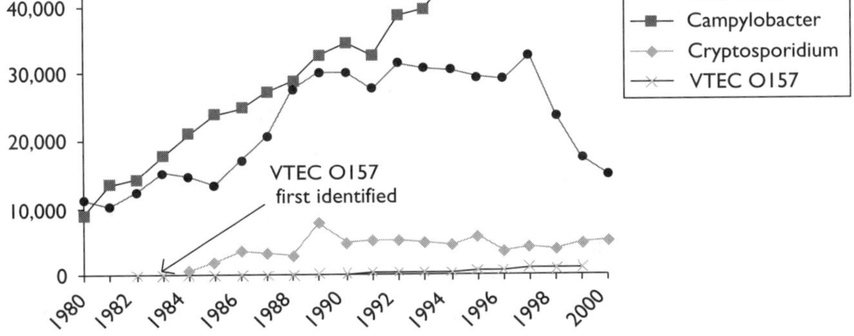

Figure 1.2 Incidence of food poisoning in England and Wales, 1980–2000
Source: Public Health Laboratory Service data (www.phls.co.uk)

The mechanisms of disease in all the strains described above depend upon genetic information, contained often on plasmids or bacteriophages or on chromosomes whose products are regulated by plasmid-encoded genes. The consequence of this is that the virulence traits can be transferred from one serotype of *E.coli* to another if they exchange genetic information on plasmids or bacteria phages. This frequently occurs in bacterial populations.

The emergence of VTEC O157

VTEC emerged in the 1980s in the United States and has since been recognised worldwide as a major cause of both self-limiting diarrhoea and more serious long-term illness, in particular, haemolytic uraemic syndrome. The way in which VTEC was recognised as a new public health problem is important to note because it teaches several epidemiological and environmental health lessons.

Two outbreaks occurred in the United States in Oregon and in Michigan in February and June 1992. A total of 47 cases were identified in people with a gastrointestinal illness which had some unusual characteristics. The main symptoms were severe colicky abdominal pain so serious that patients sometimes were admitted to hospital under the surgeons for suspected perforated bowel. They also had diarrhoea which was heavily blood stained. Normal faecal examinations did not reveal the expected bacterial infections such as Salmonella, Campylobacter or Shigella, but detailed laboratory investigations carried out at the Centers for Disease Control, Atlanta, identified an *E.coli* strain O157 from half of all the faeces specimens. Samples of faeces from well people were also examined and none of those produced this strain of *E.coli*. Epidemiological investigations linked the cases to consumption of beefburgers at a single chain of fast food restaurants. Microbiological examination of the frozen raw beef that was used in these restaurants in Michigan identified the same serogroup of *E.coli* O157. Sophisticated discrimination of the strain of the organism confirmed that the isolate from the raw meat and from clinical (patient) samples were indistinguishable (Riley *et al.* 1983).

These investigations are excellent examples of the power of epidemiological studies. They were the first outbreaks in which the clinical presentation called haemorrhagic colitis was recognised, and they identified beefburgers from fast food chains as the major vehicle of infection. Since this recognition many other similar outbreaks have been identified across the world.

The environmental health investigations were important to note. The method of contamination of the food from these fast food restaurants was carefully determined. The patty of beef recovered from the supply chain which produced the *E.coli* was of the same batch that was used in the restaurant at the time of the outbreak. The patty had never been in either of the restaurants so it could not have been cross-contaminated in the restaurants, and this ruled out an infected food handler as the principal source. Furthermore, the contaminated patty was raw, indicating the cooking process had allowed the bacterium to survive and remain in the final product. The food production process was examined carefully and it was found that the grills in the implicated restaurants had not reached the specified temperatures during busy periods.

Following these two foodborne outbreaks, in November 1982 an outbreak of diarrhoea with haemorrhagic colitis affected thirty-one out of 353 residents of a home for the elderly in Ontario, Canada. *E.coli* O157 was isolated from eighteen individuals. Epidemiological investigations showed an association between the place of eating and illness, consistent with a contaminated food stored in warming ovens before being served to residents in their rooms and in a dining room somewhere away from the kitchen. Hamburgers had been served one to two days prior to the outbreak. An important aspect of this outbreak was the biphasic epidemic curve, a feature of secondary transmission, suggesting that O157 could be passed from person to person. Since then, person-to-person transmission, especially in nursing homes and between young children, has been an important feature of the epidemiology of the disease (Stewart *et al*. 1983).

In 1982 when these outbreaks occurred it was not known how common a problem was posed by O157 and several research groups established population surveillance schemes aimed at establishing the extent of sporadic infection. The surveillance programmes soon established that sporadic cases of O157 occurred regularly, in addition to the occasional outbreaks, and that there was a wide geographical distribution in the USA, Canada and the United Kingdom. Surveillance also demonstrated that the range of clinical illnesses in those presenting with this infection was wider than the early outbreaks had suggested, and as well as haemorrhagic colitis, it included non-bloody diarrhoea, and in some cases a renal problem called haemolytic uraemic syndrome.

How VTEC causes illness in humans

Table 1.2 summarises the pathogenic mechanism of some common bacterial enteric pathogens.

To control VTEC O157, EHOs need an understanding of the ways in which it causes disease in humans. VTEC O157 possesses three potential virulence attributes: namely, 'adherence', 'attachment-and-effacement' and 'production of vero cytotoxins'.

Adherence

Microbial adherence molecules (adhesins) allow some bacteria to adhere to the gut wall by attaching to receptors possessed by the cells lining the

Table 1.2 Pathogenic mechanisms of common bacterial enteric pathogens

Ingestion of preformed toxin	*Toxin production in vivo*	*Tissue invasion*
Staphylococcus aureus (toxin stable to heat, protease, radiation and pH extreme)	*Clostridium perfringens* (toxin labile to heat and protease)	*Salmonella* (also produces cytotoxins)
Bacillus cereus (toxin stable to heat, protease and pH extreme)	*Bacillus cereus* (second toxin labile to heat and protease) *Vibrio species* VTEC O157	*Shigella* (also produces Shiga cytotoxin)

gastrointestinal tract. Attachment to the mucosal surface of the gut prevents the bacteria from being swept along by the natural peristaltic action and brings bacterial toxins into close contact with the eukaryotic cell surfaces.

VTEC O157 can adhere to the intestine of experimentally infected animals. Using animal and laboratory experiments on cell lines, some investigators have concluded that adherence is conferred by the 60MDa plasmid, possessed by all VTEC O157. Others have concluded that the adherence occurs via surface proteins (Dytoc *et al.* 1993; Karch *et al.* 1987; Levine 1987; Louie *et al.* 1993; Sherman *et al.* 1997; Tzipori *et al.* 1987; Wadolkowski *et al.* 1990).

Attachment-and-effacement

Attachment-and-effacement, though it has not been demonstrated in humans, has been described as a major morphological manifestation of VTEC O157 infection (Kelly *et al.* 1990). It has been described as an intimate bacterial attachment to the surface of enterocytes with the formation of a cup or pedestal from the cell surface, in which the bacterium sits. It has also been shown that non-O157 VTEC strains caused attaching-and-effacing lesions in experimentally infected rabbits.

Further evidence of the likelihood of attaching-and-effacing lesions in human *E.coli* O157 infection was provided by their similarity to lesions produced by EPEC. The attaching-and-effacing lesions associated with *E.coli* O157, and observed in animal experiments, were indistinguishable from the attaching-and-effacing lesions seen in humans and animals infected with EPEC strains (Moon *et al.* 1983; Tzipori *et al.* 1989).

VTEC strains cured of their vero cytotoxin associated bacteriophage and hence unable to produce VT have caused diarrhoea in monkeys (Tzipori *et al.* 1989). Also, some *E.coli* O157 isolated from humans with diarrhoea have been found not to produce VT. The diarrhoea was believed to be caused by the attaching-and-effacing mechanism.

Recent evidence has indicated that VTEC can invade certain human epithelial cell lines (T24 bladder and HCT-8 ileocecal) and that the invasion potential is chromosomally encoded (Oelschlaeger *et al.* 1994).

Toxin-production

VTEC O157 produces vero cytotoxin, so-called after the first demonstration of their toxicity on vero cells (African green monkey kidney cells). The VT produced by VTEC is indistinguishable from that produced by *Shigella dysenteriae*.

Low-level VT production has also been observed in non-pathogenic *E.coli* strains (K12) and other bacteria. An outbreak of severe gastroenteritis, HUS and TTP has been attributed to a vero cytotoxin producing strain of *Citrobacter freundii*.

It is now recognised that human isolates of VTEC O157 can, in fact, produce either or both of two vero cytotoxins which are termed VT1 and VT2.

VT1 is neutralised by antiserum against purified Shiga toxin produced by *Shigella dysenteriae* type 1. VT2 is not neutralised by the above. VT1 and VT2 are encoded by separate, identifiable bacteriophages. Variant forms of VT2, commonly referred to as VTv, have also been reported, which are not 'phage-encoded.

The toxins consist of an A (active) and multiple copies of a B (binding) subunits. The B subunit must bind to a receptor for the toxin to exert its pathogenic effect. The receptor for most VTs is a glycolipid globotriaosyl ceramide (Gb3) present on the surface of susceptible cells. The A subunit then enters the cell and inhibits protein synthesis. It is currently thought that VTs kill the epithelial cells lining the colon which leads to fluid secretion into the gut and hence to diarrhoea. Furthermore, damage of the blood vessels of the colon gives the blood seen in the faeces. The Gb3 receptor for VT is present in human renal tissue and the putative target of VT, during the development of HUS, is believed to be the renal glomerular microvascular endothelial cells. The damage caused to the renal glomerular endothelium is likely to affect haemostasis, leading to occlusion of the glomerular microvasculature by platelets and fibrin.

Various laboratory animal models have been extensively used to demonstrate the toxicity of purified vero cytotoxin on the intestine, kidneys and central nervous system. There is further evidence that VT plays a role in the pathogenesis of VTEC O157 disease. Free VT has been identified in the faeces of patients with VTEC O157 infection. Antibodies to VT have also been demonstrated in infected individuals. Furthermore, many different VTEC strains are isolated from patients with HUS indicating that the common pathogenic mechanism is production of VT.

Animal models have indicated that, where VTEC O157 infection involves a strain that produces VT2, the likelihood of HUS developing is greater although evidence as to the reasons for this are conflicting and beyond the scope of this book. Nevertheless, human studies have also shown that individuals who develop HUS are infected with a strain which produces

VT2. To summarise, VTEC O157 causes illness in humans by initially adhering to the lining of the colon and then forming an attaching-and-effacing lesion. The organism produces one or both of two vero cytotoxins. The resulting damage to the epithelial cells causes secretion of fluid into the lumen of the gut and an initially watery diarrhoea. Further damage to small blood vessels causes blood to be seen in the stool. In some individuals the VT becomes bloodborne and affects capillaries in the kidneys causing kidney failure. VT2 is more important in the development of HUS than VT1.

Incubation period

The incubation period for VTEC O157 cannot be directly ascertained from volunteer studies. It can, however, be estimated from outbreaks where a point source of infection, and hence an exact time of exposure, can be identified.

The usual incubation period is 3 or 4 days. However, incubation periods of 5–8 days are also reported and incubation periods as long as 14 days are indicated. Furthermore, incubation periods as short as 1–2 days have also been reported. Where longer incubation periods are identified, there is the possibility that the case is secondary and acquired from another case.

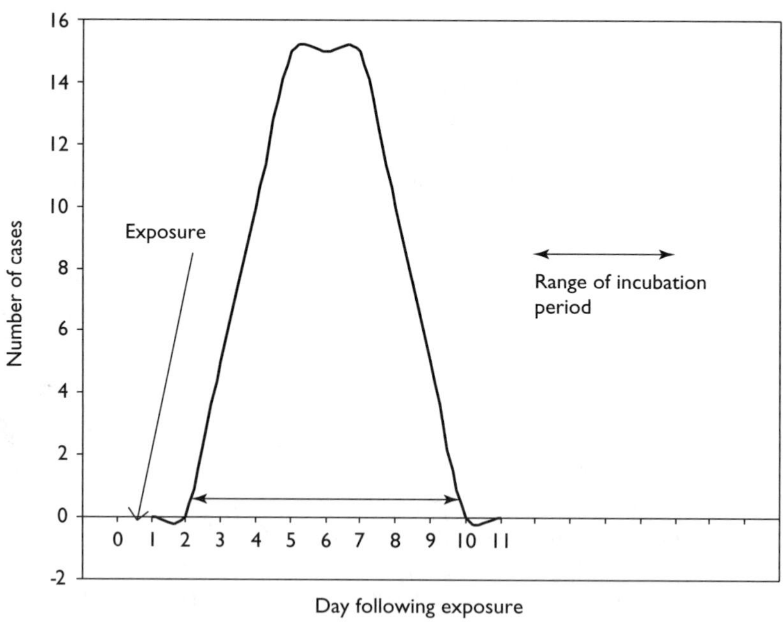

Figure 1.3 Estimation of incubation periods from outbreaks

These incubation periods are consistent with illness being caused by the attachment of the organism to the cells and toxin production in the gut rather than from ingestion of preformed toxin in foods.

Infectious dose

It is important to note that infectious dose is a population concept. There is not one single number of organisms that will make all people who are exposed ill. By analogy with the concept of 'lethal dose' in toxicology the ID 50 would be the dose that makes 50 per cent of those exposed ill. There will be an ID 10 and so on. Infectious dose will vary by person characteristics such as age and co-existing diseases, as well as by type of food consumed and other factors. Definitive volunteer studies to determine ID 50 cannot be carried out. However, the number of organisms needed to cause infection can also be inferred from studying outbreaks and sporadic cases. Person-to-person transmission of VTEC O157 is commonly reported for sporadic cases and outbreaks. During person-to-person transmission, an individual is generally exposed to only a small number of organisms, indicating a small infectious dose. Waterborne outbreaks, where the initial contamination was diluted many times, have also occurred, again indicating that very few organisms are needed to cause illness in susceptible individuals. In foodborne outbreaks, it has often not been possible to identify gross food handling errors, e.g. beefburgers have been only slightly undercooked.

Where foods have been submitted for microbiological examination, very small numbers of organisms (<2 organisms per 25 g of meat) have been found. In one outbreak associated with beefburgers, raw burgers from the implicated batch contained only 700 organisms each before they were cooked (Willshaw *et al.* 1994). In an outbreak associated with dry-cured salami, it was estimated that some cases had consumed fewer than five organisms (Tildon *et al.* 1995). Cases of laboratory acquired infection with VTEC O157 have been reported, without any identified lapses in good laboratory practice, again indicating that very few organisms are needed to cause infection. These figures are considerably less than the infectious dose of other foodborne pathogens such as *Clostridium perfringens* where more than a million organisms/g of food are required to cause illness.

Practical implications of virulence properties

In common with other food poisoning organisms, e.g. *Bacillus cereus*, VTEC O157 can produce substantial amounts of toxin in foods at 37°C but, as the toxin is destroyed by heat, this does not survive if the food is heat treated prior to consumption. Furthermore, the incubation periods are consistent with illness being caused by attachment-and-effacement and toxin production in the gut rather than a preformed toxin in foods. Therefore, in contrast to *B. cereus*, the heat treatment of food immediately prior to consumption is a critical control point. In contrast to some other food poisoning organisms, VTEC O157 does not require a period of growth in foods to reach numbers capable of causing illness, so cross-contamination does not need to be accompanied by poor temperature control for infection to occur. The prevention of cross-contamination is particularly important in preventing this infection.

The disease

Haemorrhagic colitis

The originally recognised haemorrhagic colitis typically begins with the sudden onset of severe abdominal cramps and watery non-bloody diarrhoea, followed by the appearance of blood in the stool on the second or third day of illness. Approximately half of the patients experience vomiting, and fever occurs in fewer than one-third. The illness lasts about a week, with most patients recovering without obvious sequelae; but long-term follow up of infected persons has not been reported.

Recurrent haemorrhagic colitis has been reported. There are several reports of individuals infected with VTEC being misdiagnosed as suffering from other conditions such as appendicitis, intersusception, primary inflammatory bowel disease. In some cases unnecessary surgical procedures have been performed. It has been reported that the clinical presentation of VTEC may be particularly difficult to differentiate from ischaemic colitis or pseudomembranous colitis in elderly patients.

Haemolytic Uraemic Syndrome

Although the link between O157 infection and Haemolytic Uraemic Syndrome (HUS) didn't come to light until 1981, HUS as a syndrome had been recognised for many years and is characterised by the occurrence of a rare form of anaemia, microangiopathic haemolytic anaemia, low level of blood platelets (thrombocytopaenia) and kidney failure. In most developed countries HUS is amongst the commonest causes of acute renal failure in children. The underlying pathology is thrombosis of the microcirculation, which is almost certainly a consequence of sticking of platelets in large clumps at the site of injury caused by the vero cytotoxins. The link with diarrhoeal illnesses had been recognised for some time and many cases of HUS occurred in patients who reported a severe diarrhoeal illness a week or so before onset of the renal failure. However, it wasn't until the discovery of *E.coli* O157 that the main cause of this renal disease was recognised.

HUS is an important component of the public health problem posed by VTEC O157 because the overall case fatality rate is between 6 and 10 per cent. Up to 40 per cent of patients followed up for several years were thought to have long-term abnormalities of renal function. The link with VTEC infection may often be missed if patients do not consult their doctors for the initial diarrhoeal illness; they may not be excreting VTEC when they present a week or two later with the symptoms of HUS.

Non-bloody diarrhoea

Typically the illness begins with a sudden onset of severe abdominal cramps and watery non-bloody diarrhoea. About half of the patients present with vomiting, but fever occurs in fewer than a third: it is more common in those with severe illness. Typically this lasts about a week, with most patients recovering with no obvious side effects. VTEC O157 has been isolated from patients with chronic diarrhoea but it is uncertain at this stage whether this is the cause.

Other diseases

Thrombotic thrombocytopaenic purpura has been reported as have rare cases of bowel perforation, bowel necrosis, toxic mega-colon, gastrointestinal tract stricture, pancreatitis, hepatitis and neurological abnormalities. Extra intestinal isolations have been reported in patients who also had diarrhoea.

The more severe forms of VTEC infection – haemorrhagic colitis and HUS – do not represent the full spectrum of the disease in the community. These more severe presentations are more likely to be recognised and

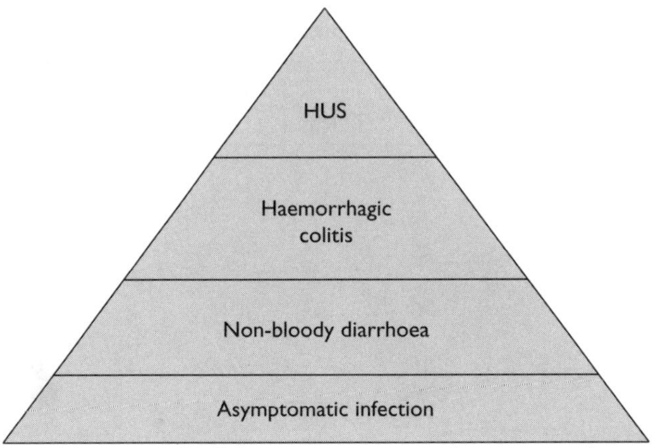

Figure 1.4 Pyramid of spectrum of disease caused by VTEC O157

investigated because of their severity. Information about the full spectrum of the presentation can be obtained from surveillance and outbreak investigations. Surveillance data can, however, lead to biases because of particular policies for screening used by participating laboratories. For example, some laboratories only examine faecal samples from cases with bloody diarrhoea for VTEC infection and therefore non-bloody diarrhoeal cases will not be represented in their data. In the Welsh population surveillance system in which all faecal samples were examined by standard methods for VTEC it was found that 82 per cent of cases had diarrhoea, 62 per cent had abdominal pain, 46 per cent had blood in their stools, 41 per cent had the classical haemorrhagic symptoms, and 4 per cent developed HUS.

Further information on the spectrum of illness has been obtained from point source outbreaks in definable cohorts. For VTEC such outbreaks have occurred mainly in children's day-care centres and in old people's homes, hence the populations have been children or the elderly and are only applicable to those sub-populations. Table 1.3 provides data from selected outbreaks in which the cohorts affected were defined and all cases of infection could be identified. The studies show consistently that up to 25 per cent of infections may be asymptomatic and that only about a half of patients affected with symptoms develop bloody diarrhoea.

The proportion presenting with HUS is variable and depends very much on the susceptibility of the group: for example, young children aged 1 to 4 seem at a much higher risk. Case fatality rates vary but can be very high in the elderly. The duration of illness varies with the age of patients. Children in one study suffered illnesses on average of 9 days compared to 6 days in adults. It has been suggested in some studies that patients given anti-diarrhoeal medication were more likely to develop HUS; other studies have found an association between taking antibiotics and prolonged

Table 1.3 Severity of illness in VTEC O157 infection – outbreak reports

Population	*Setting*	*Asymptomatic (%)*	*Bloody diarrhoea (%)*	*Hospitalised (%)*	*HUS (%)*	*Died (%)*
Paediatric[1]	Day care	17	50		17	
Paediatric[2]	Day care				7	
Paediatric[3]	Day care	41	41		18	0
Elderly[4]	Nursing home		75		24	35
Elderly[5]	Nursing home		56			12
Community[6]	Community		81	82	31	0
Community[7]	Christening party cohort		44		5	0

[1] Belongia *et al.*, 1993; [2] Spika *et al.*, 1986; [3] Allaby and Mayon-White, 1995; [4] Carter *et al.*, 1987; [5] Ryan *et al.*, 1986; [6] Morgan *et al.*, 1993; [7] Salmon *et al.*, 1989.

bloody diarrhoea. However, the results of different studies are inconsistent, some suggesting that antibiotics may lessen the likelihood of developing HUS.

Factors influencing the emergence of VTEC

Evidence suggests that the reservoir of VTEC O157 is cattle and that humans become infected with VTEC when they ingest the bacteria contaminating food or water, or by direct contact with infected animals or the contaminated environment. But how new is VTEC?

There are no reports of outbreaks of haemorrhagic colitis before 1982, and no evidence that *E.coli* O157 had been a cause of any outbreaks in the UK since 1973. It is unlikely that outbreaks of such severity would have remained unidentified, hence it is thought that haemorrhagic colitis due to VTEC is indeed a new phenomenon. On the other hand, outbreaks of haemolytic uraemic syndrome were reported in the 1960s. The situation regarding sporadic infections is less clear, although there is some evidence to suggest that VTEC was identified, albeit very rarely, prior to 1982. Laboratories in the USA and Canada and UK have retrospectively reviewed the serotypes of isolates of *E.coli* sent to them since 1973. Amongst the 3,000 specimens received by CDC, they found a single isolate of O157 – a Californian woman with grossly bloody diarrhoea (Riley *et al.* 1983). In the UK, over 15,000 strains from 1978 to 1983 were reviewed by PHLS, and there were 28 O157 isolates (Day *et al.* 1983). Only two of the strains produced VTEC and only one was isolated from a patient who had suffered diarrhoea. In Canada, the Laboratory Centre for Disease Control found O157 in six of 2,000 isolates sent to them over the same period (Johnson *et al.* 1983). Collectively these data do indicate that VTEC cases were identified prior to 1982 but only in very small numbers. We cannot therefore discuss the emergence of VTEC as an artefact of surveillance, but reasons for the recent emergence of VTEC need debate.

Changes in the livestock industry

In the UK and the USA, small family farms have to a large extent been replaced by large-scale centralised processing. Average dairy and beef cattle herd sizes increased significantly in the 1980s. For example, average herd sizes in Wales rose from thirty in 1974 to forty-eight in 1987. In the USA the number of herds with less than twenty head of cattle fell by 23 per cent from 1978 to 1987. A new ecology with crowded animal populations has been created increasing the probability of spread of bacteria from one infected animal to another. There have also been changes in the food industry which may have increased the risk of VTEC.

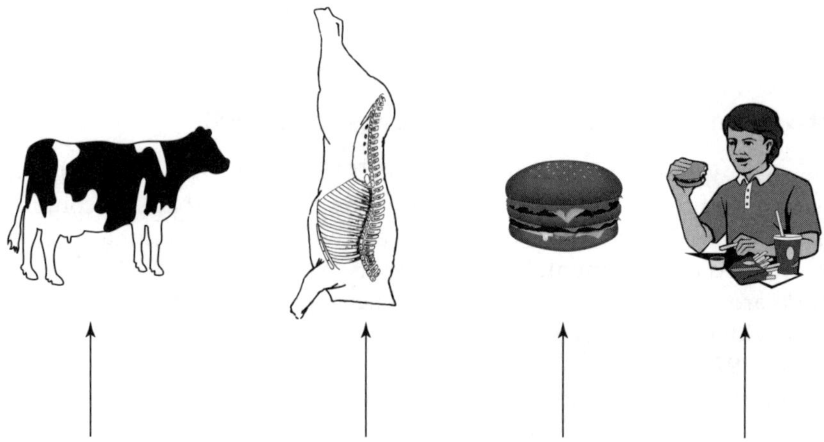

Figure 1.5 Factors influencing emergence of VTEC O157

Changes in food production

The production of food has become more centralised. Therefore, contaminated meat and milk products from one animal can cross-contaminate many more and affect potentially huge numbers of the population virtually simultaneously. Hence, food products with only very low levels of contamination take on as much public health importance. This is demonstrated by a huge outbreak in the USA in 1992/3, where as many as 500 cases of VTEC O157 were associated with the consumption of contaminated hamburgers from a single restaurant chain. At least five slaughter plants in the USA and one in Canada had supplied the meat for one day's production of beefburger patties (CDC 1993). It should be noted, however, that such outbreaks, affecting individuals spread over thousands of miles and possibly several countries, are notoriously difficult to detect unless sound surveillance methods are in place. The few cases that are seen locally will not be distinguishable for the background level cases. There have also been changes in consumption patterns.

Consumer behaviour

In the UK, the consumer catering market grew by 63 per cent, between 1986 and 1991 to £16 billion, with price inflation of only 36 per cent over the same time span. Furthermore, over the same period, the number of meals eaten outside the home increased by about 10 per cent. The population, therefore, is increasingly exposed to foods which are mass-prepared, with the potential for large food poisoning outbreaks should there be a food handling error.

Social behaviour

Social changes have occurred which give increased opportunities for person-to-person spread to occur in susceptible groups. In the UK, the growing sub-group of elderly people (4.7 per cent of the population were over 75 in 1971 compared to 7.0 per cent in 1991) means that the population is increasingly immunocompromised by age and hence more vulnerable to enteric pathogens. A substantial proportion of the elderly (about 20 per cent of the over 85-year-olds in 1991) are now living in group homes and institutions. In Wales, the number of elderly people in homes registered under the Registered Homes Act 1984, increased from 9,076 in 1980 to 13,608 in 1991. Such homes have been the setting for point-source (often foodborne) outbreaks of VTEC O157 with subsequent person-to-person transmission. Case fatality rates for VTEC O157 outbreaks in such settings have been as high as 35 per cent.

Child care is also increasingly institutionalised. In 1993 there were about 985,000 day-care places for children under 5 years in the UK, compared with only 409,000 in 1971. Such settings increase the number of children with whom a given child has contact. Furthermore, children (particularly those in nappies) cannot exercise careful personal hygiene. The combination of these two factors results in opportunities for considerable person-to-person spread of enteric pathogens with low infectious doses. It is estimated that only 10–100 organisms are needed to cause VTEC O157 infection. Furthermore, the under 5's are particularly likely to develop the serious sequelae associated with VTEC O157 and outbreaks in day-care centres are a serious public health problem. Increasingly mothers who are working outside the home and taking time off work to look after a child may be a problem. This increases pressure for the child to stay in nursery and thereby increasing the risk of transmission. Policies for excluding children with diarrhoea and with confirmed VTEC O157 from day care are discussed below.

International travel

Finally, there has been a huge increase in travel which has aided the rapid dissemination of pathogens between countries. For example, in 1971 just over 25 million international air passengers passed through UK airports, but by 1994, this had increased to just under 100 million. The associated international transportation of goods (including animals) provides another route through which diseases arising in one country can be introduced to suitable habitats (if they exist) in others.

Summary

- *E.coli* O157 is not the only *E.coli* which causes diarrhoeal disease.
- It is sometimes referred to as an enterohaemorrhagic *E.coli* or EHEC.
- *E.coli* O157 was first identified as a pathogen in the USA in 1982.
- It is likely to have emerged some years before its discovery as a result of changes in the livestock industry, food production, consumer and social behaviour.
- It causes a spectrum of disease ranging from mild diarrhoea, haemorrhagic colitis (diarrhoea, abdominal pain and blood in stool) to haemolytic uraemic syndrome.
- The incubation period is probably 3–4 days but the range is likely to be 1–14 days.
- It causes illness by adhering to the intestine wall, forming an attachment and effacement lesion and producing one or both of two vero cytotoxins.
- The infectious dose is very low, and possibly as few as ten organisms can cause disease in some individuals. Consequently, person-to-person spread is a real possibility and much more likely to occur than with similar organisms, e.g. Salmonella.
- The low infectious dose means that the bacteria does not necessarily need a period of multiplication on food to reach an infectious dose. Therefore, preventing the initial cross-contamination of cooked foods is even more important.

KEY MESSAGES

- VTEC O157 is a rare but serious disease with a low infectious dose.
- It emerged sometime prior to 1982, as a result of changes in animal rearing, food production, consumer and social behaviour.

Chapter 2

How is VTEC diagnosed?

VTEC O157 should be considered as a suspect food poisoning and therefore must be reported to the Proper Officer of the local authority under the Public Health (Control of Disease) Act 1984. Presumptive isolates of VTEC O157 should be immediately reported by phone or fax to the relevant public health authority for investigation.

Examination of faecal specimens

The examination of faecal specimens for VTEC O157 can follow either of two approaches. The first attempts to isolate *E.coli* of serogroup O157 and then confirm whether the isolate produces verotoxins. The second approach initially identifies any evidence of verotoxin production and then attempts to isolate and identify the organism involved.

A feature of the presentation of illness in the earliest outbreaks was that unless the faecal samples were obtained within 5 days of onset then the chances of the sample being positive were very small. This drew attention to the importance of the culture media used by laboratories. Considerable work has been done on culture methods since then to increase the sensitivity of the media.

Isolating *E.coli* O157

This method is used routinely in local microbiology laboratories and cases reported to Environmental Health Officers (EHOs) are most likely to have been identified in this way.

VTEC O157:H7 is a gram negative bacillus. It is similar to most *E.coli* in its ability to ferment lactose. However, unlike approximately 80 per cent to 90 per cent of *E.coli* in human faeces, it does not ferment sorbitol within 24 hours. This characteristic (a biochemical marker) has been of great use in the development of diagnostic laboratory methods, although it is not an absolutely reliable test since there are rare strains of VTEC O157 that do ferment sorbitol.

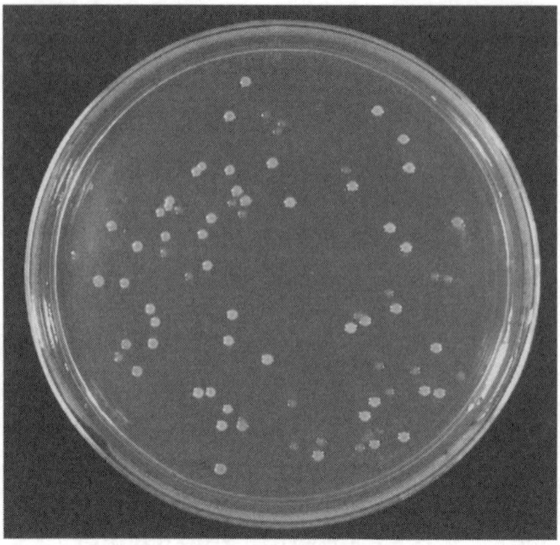

Figure 2.1 Sorbitol MacConkey Agar with O157 colonies

Faecal specimens are usually examined for *E.coli* O157 by culture on Sorbitol MacConkey Agar (SMA) in which the lactose has been replaced by 1 per cent D-sorbitol and which is commercially available. The non-sorbitol fermenting *E.coli* O157 colonies appear as small, round, smooth, greyish colonies which are then tested for agglutination with *E.coli* O157 antiserum or with an O157 agglutination kit. Some investigators also test for the H7 antigen either with antisera or by motility inhibition tests.

Biochemically VTEC O157 behaves as a typical *E.coli* except for failing to ferment sorbitol or produce B-glucuronidase. Biochemical tests can confirm that the colonies are not species such as *Brucella abortus*, *Salmonella* group N, *Vibrio cholerae*, *Yersinia enterocolitica* O9, which may cross-react with the O157 antiserum.

The use of Sorbital MacConkey as a screening medium has been widely accepted. However, several other *E.coli* serogroups are non-sorbitol fermenters, as are other species found frequently in human faeces such as *Proteus* spp. To overcome this, cefixime and rhamnose can be added to SMA which is referred to as CR-SMAC. VTEC O157 do not ferment rhamnose on agar plates in 24 hours, whereas many (60 per cent) of the other non-sorbitol fermenting *E.coli* do. Furthermore, cefixime (a cephalosporin antibiotic) is more active against *Proteus* spp. than *E.coli*. The PHLS standard operating procedure (SOP) specifies direct inoculation of faeces onto CR-SMAC. However, this method may not detect *E.coli* O157 in specimens obtained over 4 days after onset of symptoms or after HUS has developed. In these cases the number of bacteria in the stool is likely to be low. This is also true for faecal samples of contacts in family/general outbreaks and in convalescent specimens. In these cases, enrichment of faecal specimens in modified tryptone soya broth increases detection in primary cases by 15 per cent and in contact and convalescents by 30 per cent. Innoculated plates are incubated at 37°C for 18 hours.

Immuno-magnetic separation (IMS) can also be used for convalescent and contact specimens. This technique is commonly used to isolate O157 from foods and milk. It uses paramagnetic particles coupled to specific antibodies to the O157 antigen to capture organisms from an enriched culture. Bacteria adhering to the beads are separated by a magnetic field, washed and plated on CR-SMAC.

Identifying toxin production

VTEC O157 produces vero cytotoxin, so-called after the first demonstration of toxicity on African green monkey kidney cells. The verotoxins produced are indistinguishable from those produced from *Shigella dysentariae*. Verotoxins fall into two categories, either or both or which can be excreted by isolates. They are termed VT1 and VT2. VT1 is neutralised by antiserum against purified toxin produced by *Shigella dysenteriae* type 1. VT2 is not neutralised. VT1 and VT2 are encoded by separate identifiable bacteria phages. These toxins bind to receptors on the surface of susceptible cells. Part of the toxin enters the cell and inhibits protein synthesis. Consequently epithelial cells lining the colon are killed and fluid is secreted into the gut. Damage to blood vessels in the colon gives rise to blood in the faeces. Verotoxin attacks kidney tissue and damages the tiny blood vessels in the glomerulus.

Advantages

- Routine method used by most microbiology laboratories and currently recommended by PHLS.
- Relatively easy and quick to perform and therefore available in all laboratories.
- Fairly reliable.

Disadvantages

- Will not identify VTEC O157 which ferment sorbitol or which do not express the O157 antigen.
- False-positives can rarely occur.
- Not as sensitive as some more sophisticated methods, particularly for stools collected more than 4 days after onset, convalescent and contact samples.

Figure 2.2 CR-SMAC – advantages and disadvantages

The strains to be tested are usually grown in Trypticase soya broth or in Penassay medium, and culture filtrates are added to monolayers of vero cells. Cells become rounded and detached in the presence of VT. Final readings are usually made after incubation for 3 to 4 days when the cells are fixed and stained. Either individual colonies or broth inoculated with sweeps can be examined. Confirmation that the observed cytotoxic effect is actually caused by VT can be made by neutralisation tests using antisera against VT1 and VT2. The heat-lability of the toxin can be confirmed by heating samples at 100°C for 15 minutes before testing for VT activity as above. These methods have been widely used to detect free VT in stools. The major disadvantage of the vero cell assay is the 3-day turnaround time.

VT can also be detected using an enzyme linked immunosorbent assay (ELISA). Antibodies are used to bind the toxin which is then detected with monoclonal or polyclonal antiserum against VT. The ELISAs are not as sensitive as the vero cell test but are easier to perform. VT toxins show variation in their antigenicity and binding, therefore care must be taken in the choice of reagents if all VT producing strains are to be detected. The advantages of an ELISA are that it has a shorter turnaround time than DNA hybridisation methods (see below). VT positive samples need to be processed using an immuno-assay, cell culture or DNA hybridisation method to detect and isolate individual colonies.

The disadvantages of cell culture assay to detect free VT are that it can take 3 to 7 days to obtain a result. Furthermore, the process is prone to contamination especially when examining faecal supernatants as a number of other microbial toxins, e.g. *Clostridium perfringens* enterotoxin, *Clostridium difficile* cytotoxin and *E.coli* LT, may be present in the faecal sample, and have an effect on the vero cells. To overcome this, a neutralisation test can be carried out with antitoxins against VT, but this is not yet commercially available in the UK. Cell culture assay cannot be carried out in laboratories that lack tissue-culture facilities, and the sensitivity of tissue-culture cells decreases over time, causing assay to assay variability.

Once several investigators had successfully cloned the genes specifying VT1 and VT2, DNA probes specific for VT genes were developed that could

Advantages

- Sensitive and simple to perform.

Disadvantages

- Labour intensive and impracticable for routine use.
- False-positives can occur due to contamination.

Figure 2.3 IMS – advantages and disadvantages

detect all VTEC in colony hybridisation tests. One type of probe is a polynucleotide probe for VT1 and VT2, which is able to detect VTEC present in low numbers. These probes have been widely used for several years in the diagnostic setting at the Central Public Health Laboratory (CPHL) in London. The DNA probe test can be performed using membranes with faecal blots. Alternatively, a large number of colonies from a sample, or purified *E.coli*, can be examined by probes for VT1 and VT2. They are at least as sensitive as methods to detect faecal VT and more sensitive than culture on SMA.

In a variation of this approach, DNA sequences for VT genes can also be detected by amplification in a polymerase chain reaction (PCR). PCR tests can be performed on purified strains or preparations from faecal specimens. Recently, Cebula *et al.* (1994) described a PCR method that can simultaneously identify *E.coli* of serotype O157 and whether VT1 and/or VT2 genes are present. PCR can detect VT1 and VT2 in primary faecal cultures and is capable of detecting as few as 10 VT-producing organisms in a background of 10^9 other organisms per millilitre. A disadvantage of PCR is that the detection of a specific gene in a bacterial genome does not indicate whether that gene will be expressed in a biologically active form; nor does it indicate the level at which expression occurs. Furthermore, it is technically demanding, expensive and prone to false-positive results due to contamination.

Strain discrimination

VTEC O157 can be divided into a number of subtypes by a range of strain discrimination techniques. This is particularly useful for the detection of outbreaks. In addition, strain discrimination techniques can be used to link particular strains of bacteria with particular reservoirs of infection. For example, in the case of *Salmonella enteriditis* 'phage type 4, a poultry reservoir has been identified. Various schemes for discriminating strains of VTEC O157 are considered below.

Phenotypic typing is achieved by 'phage typing and identification of VT type. A 'phage-typing scheme, originally developed in Canada in 1987 (Ahmed *et al.* 1987; Khakhria *et al.* 1990), has been refined by the PHLS, Laboratory of Enteric Pathogens (LEP), and recognises over eighty types (Frost *et al.* 1989, 1993).

Of the 485 isolates 'phage-typed from Scottish samples between 1984 and 1993, 33 per cent were 'phage type (PT) 2 and 36 per cent were PT49 (Davis and Brogan 1995), although PT4 and PT48 were common in 1994 because of their involvement in three outbreaks (Coia *et al.* 1995).

Between 1989 and 1991, 36 per cent of isolates examined in the UK were PT2 and 29 per cent were PT49, 10 per cent were PT1 and 9 per cent were PT4 although these data include isolates from Scotland (Frost *et al.*

1993; Thomas *et al.* 1993). Considering isolates received between 1992 and 1994, which did not include Scottish isolates, PT2 accounted for 46 per cent, with PT49 (17 per cent), PT8 (8 per cent), PT1 (7 per cent), PT4 (6 per cent) and PT14 (4 per cent) also common. There were no marked geographical variation in 'phage type (Frost *et al.* 1989).

VT typing by gene probe can also be used to differentiate strains of VTEC O157. A single strain can produce either VT1 or VT2 or both. Several studies have shown that the strains of VTEC O157 isolated from patients with HUS almost always produce VT2, with or without VT1. The commonest VT type between 1989 and 1994 and between 1995 and 1998 for the UK as a whole was VT2 (about 79 per cent) (Thomas *et al.* 1993, 1996). There are further techniques such as biotyping by biochemical tests and antimicrobial susceptibility tests.

Plasmid analysis can also be used to differentiate strains. Multilocus enzyme electrophoresis, DNA fingerprinting and detection of restriction fragment length polymorphisms by hybridisation with lambda-phage DNA are useful in differentiating strains in epidemiological studies. In restriction fragment length polymorphism (RFLP), bacterial DNA is digested into fragments. The fragments are separated by agarose gel electrophoresis and then the fragments containing specific sequences are detected using a labelled piece of homologous DNA as a probe.

In the UK, DNA-based examination of isolates with the same 'phage and VT type have been successfully used to trace additional cases and estimate spread of infection in outbreaks. Pulsed field gel electrophoresis (PFGE) is a variation of agarose gel electrophoresis in which the orientation of the electric field across the gel is changed periodically (pulsed) rather than being kept constant. Consequently, even megabase DNA fragments can be effectively separated by size. In practice, during the investigation of sources and outbreaks a combination of the above techniques is used, including pulsed field gel electrophoresis. Research is underway to assess whether it is cost-effective to operate subtype specific surveillance for identifying VTEC O157 associated outbreaks. Early results indicate that if five cases are averted annually the system would recover costs.

Serology

Serology can be used to detect antibodies to O157 lipopolysaccharide (LPS) or to detect antibodies to verocytotoxins. Several classes of immunoglobulin response can be measured. Age does not appear to affect the antibody response. Cross-reactions have been demonstrated between the *E.coli* O157 LPS and the O antigen of other bacteria. Hence, false-positive antibody titres have been reported in patients with yersiniosis and brucellosis. Strains of *Vibrio cholerae* O1 Inaba and group N *Salmonella* share epitopes with *E.coli* O157 and cross-reactions have been observed with some strains of *E.hermanni* and *Citrobacter freundii*. The duration of the O157 antibody response will affect case finding in outbreaks and diagnosis of HUS. Studies in the USA have shown that mean antibody titres to O157 LPS declined to near negative by 8 weeks after the onset of diarrhoea, although some patients had positive titres at 7 months. IgA antibody titres declined to near negative by 4 weeks. Positive titres to IgG peak in the second or third week of illness (Barrett *et al.* 1991; Bitzan *et al.* 1993).

VT has never been detected in serum from patients with HUS or haemorrhagic colitis. In early studies of VTEC infection, investigators demonstrated that patients developed rising titres of VT neutralising antibodies and used this to diagnose VTEC infection where faecal cultures were negative. Barrett *et al.* (1991) investigated the production of antibody titres to VT. They found that a small proportion of patients infected with VT1-producing *E.coli* O157:H7, produced antibody titres to VT1 but none produced antibodies to VT2. In an outbreak of a strain of VTEC O157, which was demonstrated to produce VT1 and VT2, an ELISA was used to detect VT1 antibodies in eleven out of forty-seven (23 per cent) of the patients. Positive titres were also recorded in 11 per cent of well controls, but most titres were lower than those recorded for patients. In the same study, none out of eighty-three cases in two outbreaks of VTEC O157 developed antibodies to VT2 as detected by neutralisation or ELISA. The first outbreak used in the investigation was caused by a strain that produced VT2 only and the strain involved in the second outbreak produced both VT1 and VT2. Sera from all patients and controls exhibited non-specific neutralisation of VT2. In a study in the UK, sera from patients known to be infected with a VTEC strain that produced VT1 and VT2, did not have VT1 neutralising activity although two sera from healthy adults did neutralise VT1. In contrast, VT2 neutralising activity was found in most of the sera from patients and all of the sera from healthy adults (Scotland *et al.* 1994). Salivary tests for seroconversion are being developed which will be less invasive than the previously described tests which require blood samples to be taken. Serodiagnosis is available at LEP and the Scottish *E.coli* O157 reference laboratories in Aberdeen.

Period of excretion

Following onset of the acute diarrhoeal symptoms associated with gastroenteric pathogens, the patient frequently continues to excrete the organism in diminishing numbers in their faeces for a period of time, i.e. the 'period of excretion'. If a stool sample is collected too long after onset, the numbers of bacteria will be so low that infection will not be detected. If the usual period of excretion for VTEC O157 is known, then if a negative result is obtained, an EHO can interpret whether it genuinely indicates that infection with VTEC O157 has not occurred or if the sample was taken too long after onset to isolate VTEC O157. This is particularly useful during the investigation of outbreaks.

It has already been shown that the infectious dose for VTEC O157 is very low and that it therefore spreads readily from person-to-person via the faecal-oral route. Secondary transmission of VTEC O157 in homes and in institutional care premises, such as children's day-care centres and nursing homes, is a particular risk. Whilst the convalescent case is still excreting VTEC O157 they are a potential source of infection to others. The effective control of person-to-person spread relies on data indicating the length of time over which the infected individual will excrete the bacteria. Data on the period of excretion of VTEC O157 are therefore particularly relevant to the environmental health control of the disease (Table 2.1).

Pai *et al.* (1988) serotyped ten randomly picked lactose-positive colonies in several stool specimens obtained from cases. They found that, up to the

Table 2.1 Period of excretion of VTEC O157

Riley *et al.* (1983)	VTEC isolated from 9/12 stools collected within 4 days of onset and from none collected 7 days or more after illness (p = 0.002, Fishers exact 1 tailed).
Remis *et al.* (1984)	24/28 (86%) haemorrhagic colitis patients with culture confirmed VTEC O157 had submitted faeces within 6 days compared to 25/45 (55%) unconfirmed (p = 0.01, Fishers exact 2 tailed).
Stewart *et al.* (1983)	VTEC O157 was isolated from patients' stools on first day of onset up to 11 days after onset. In 2/18 (11%) VTEC O157 was isolated 1–2 days after the diarrhoea had resolved.
Pai *et al.* (1988)	Period of excretion* in 63 index cases ranged from 4 to >43 days; 50% were culture negative at 10 days after onset; 42 patients were studied for more than 3 weeks or until 3 consecutive negative cultures were obtained and 11 (26.2%) were positive for more than three weeks.
	Stool cultures were positive for more than three weeks in 9/17 (52%) children under 4 years compared with only 2/25 (8%) older children or adults (p = 0.004 2 tailed fisher exact test).
	Stool specimens were collected from culture positive family members at 4.2 ± 7.5 days (mean ± SD) compared to 10.6 ± 4.4 days for culture negative family members (p<0.05, students t test).
Belongia *et al.* (1993)	24 children studied; mean duration of shedding = 17 days, one child treated with amoxicillin excreted for 62 days.
Karch *et al.* (1995)	53 children studied; median duration of shedding = 13 days, one child = 124 days.

* Period of excretion = period from onset of symptoms to first negative stool specimen.

fourth day after illness, *E.coli* O157 was identified in more than 5/10 (50 per cent) of colonies from the majority of specimens. Five stools collected from 2 adults more than 5 days after illness were negative, whereas, 10/10 (100 per cent) of colonies from 4 children collected between 7 and 13 days after onset were positive. Similarly, stool specimens collected within 5 days after onset showed high cytotoxin titres (indicating high levels of cytotoxin) whereas most stools collected after that date had low or no cytotoxic activities. They concluded that whilst most adults cease to shed the bacteria after 8 days, children shed up to 14 days and possibly longer.

There is some epidemiological evidence to suggest that prolonged excretion can occur. VTEC O157 has been isolated from an individual who had suffered chronic diarrhoea for 1 year suggesting that VTEC O157 may be excreted for this length of time (MacDonald *et al.* 1988). Pai *et al.* (1984) reported a 3-year-old female who developed laboratory confirmed VTEC O157 infection about 2 months after 4 members of her family had had a diarrhoeal illness of unknown aetiology following a hamburger meal at a fast food restaurant. Two of the 4 family members were found to have VTEC O157 in their stools. Possibly however, the initial family illness could have been unrelated and the 3-year-old female transmitted the infection to her family who experienced asymptomatic infection. Reports also indicate that children can excrete the organism intermittently and without symptoms.

Future developments

Research is underway to further improve and refine isolation techniques for *E.coli* O157. In particular, rapid detection kits are being developed that can detect VTEC O157 directly from faeces in 8 hours or less. Some are portable and can be used 'in the field' without the need to transport samples to a laboratory and without the need for culture. However, those techniques are still under development and not available for general use.

Further work on non-invasive diagnostic tests, e.g. salivary based tests, sophisticated subtyping techniques bases on DNA analysis, are also being developed further.

Summary

- Routine diagnosis is by isolation of *E.coli* O157 on Sorbitol MacConkey Agar. O157 does not ferment sorbital and colonies can be identified.
- Strains of O157 can be identified by use of 'phage types or gene probes.
- Serology may be useful when isolation has failed or is not possible.
- Infected adult patients usually excrete VTEC for less than a week. In children, the excretion period is often longer.

- Ensure that isolates are forwarded to a reference laboratory for confirmation and that the confirmation results are also forwarded to you.

Suspect cases

- Samples may be submitted during investigation of a 'suspect food poisoning'.
- If the sample is collected more than 4 days after onset of symptoms discuss the possibility of prior enrichment with your laboratory.
- If the sample is negative but VTEC O157 is highly suspected on clinical grounds (HC, HUS) discuss the following diagnostic tests with the laboratory.
  - Forwarding sample to a reference laboratory for tests for VT production: the strain could be a sorbital fermenting VTEC O157 or may not be expressing the O157 antigen.
  - Could a VT producing *E.coli* other than *E.coli* O157 be implicated? Forward sample to reference laboratory.
  - Serum or saliva serology.

Convalescent/Contact samples

- Could enrichment be used?
- Could IMS be used?
- Presumptive positive faecal specimens should always be assumed to be positive until proved otherwise.
- Negative faecal samples do not necessarily mean no VTEC O157 are present. Ask questions about further tests.

KEY MESSAGES

- Ensure that your local microbiology laboratories routinely screen all diarrhoeal stools for VTEC O157 using the recommended methods.
- Ensure that presumptive VTEC O157 isolates are reported to you by fax or telephone immediately before the sample is forwarded to a reference.

Chapter 3

How common is VTEC infection?

Data on the incidence of VTEC comes from two main sources. First, population based surveillance, and second, special studies.

Population based surveillance

Surveillance has been defined as the continued watchfulness over the distribution and trends of incidence through the systematic collection, consolidation and evaluation of morbidity and mortality reports and other relevant data. To this should be added the important step of feeding back information to those who should take control actions and to those providing data. Surveillance should provide 'information for action'.

The objectives of population surveillance are:

1 To describe the basic epidemiology of VTEC such as its geographical spread and its age incidence, in order to develop hypotheses about causation which can then be tested by research.
2 Early detection of clusters and outbreaks or new exposures to enable rapid investigation and application of appropriate control measures.
3 Monitoring trends in disease and serotypes of the organism and risk factors in order to assess priorities for intervention (e.g. open farms, burger vans, butchers' shops), research and to evaluate disease-control measures and preventive programmes.

The main steps of surveillance are:

- systematic collection of data
- analysis of these data to produce statistics
- interpretation of the statistics to provide information
- distribution of this information to all those who require it so that action can be taken, and to those providing data
- continuing surveillance to evaluate the action.

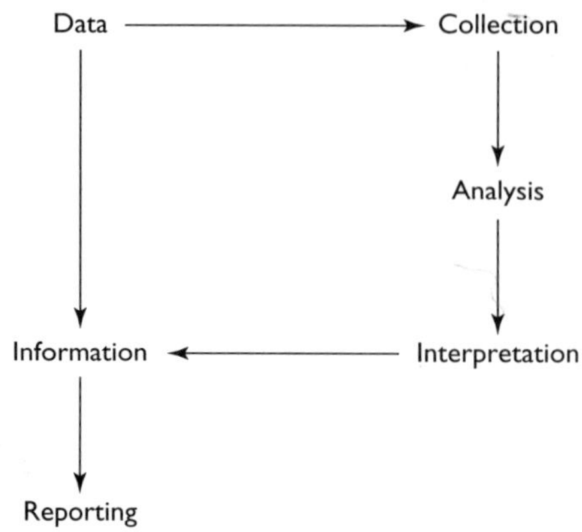

Figure 3.1 Surveillance flow of information

Key attributes of a surveillance system are simplicity, flexibility, acceptability, sensitivity, specificity, representativeness, timeliness and usefulness.

Sources of surveillance data

Statutory notifications

All doctors diagnosing certain diseases have a duty by statute to report to the Proper Officer of the local authority so that suspected cases can be investigated and control measures taken as appropriate. 'Food poisoning' is a notifiable disease although the bacterial causes of food poisoning are not in themselves notifiable. This has led to some confusion. The Advisory Committee on the Microbiology Safety of Food proposed a working definition of food poisoning which was accepted by government departments and circulated to all doctors in 1992. Its definition of food poisoning is 'any disease of an infectious or toxic nature thought to be caused by the consumption of food or water'. This definition has previously been adopted by the World Health Organisation.

VTEC disease is quite likely to be foodborne even though person-to-person spread is a significant problem. Therefore suspected and confirmed VTEC infection should be notified as a suspected foodborne disease.

The purpose of notification is to alert local authorities to take appropriate action to investigate and control incidents; data for surveillance is a secondary benefit. However, it is generally accepted that food poisoning notifications are a poor indication of the true incidence of food poisoning. This is because people with the symptoms of food poisoning often do not visit their General Practitioners. When a GP does see a patient with foodborne disease it is usually not possible to distinguish this from other more common causes of gastroenteritis. In individual cases the food history is seldom helpful. Only when a faeces sample yields a foodborne pathogen is a measure of confidence obtained. However, from a clinical point of view there is usually little if any benefit to the patient from the results of a faeces sample and therefore GPs do not routinely send samples to the laboratory. When they do it tends to be because of severity or persistence of symptoms. When VTEC causes acute haemorrhagic colitis the symptoms will certainly stand out from the normal cases seen by a GP, but unless there is a recognised outbreak in the area levels of awareness may not be high. Other more common pathogens such as Campylobacter and Salmonella and Shigella can cause bloody diarrhoea.

Laboratory reports

Reports by diagnostic laboratories provide the backbone of infectious disease surveillance in the UK. In England and Wales the Public Health Laboratory Service receives reports from over 200 clinical and public health laboratories and in addition uses the data held by the PHLS Laboratory of Enteric Pathogens which is the national reference laboratory for VTEC infections. In Scotland, SCIEH receives laboratory reports. Although these data are highly specific because they are based on the most comprehensive diagnostic tests, the number of reports received represents only a small fraction of the true incidence of these pathogens both because of the low level of microbiological investigation in general practice and incomplete laboratory reporting. Information accompanying these reports is limited and there are no data on the outcome of infected patients. The data do not include patient identifiable details so that the area of residence cannot be mapped.

Outbreak reports

In addition to the reporting of individual cases national surveillance centres receive reports of cases of VTEC outbreaks in the community. A general outbreak is defined as 'affecting members of more than one private residence or residence of an institution'. General outbreaks are distinguished from family outbreaks affecting members of the same private residence only. In 1992 an improved systematic surveillance system for outbreaks was developed by the PHLS in England and Wales in response to the Richmond's Committee recommendations. The system works in the following way. The Communicable Disease Surveillance Centre (CDSC) is made aware of outbreaks from laboratory reports, Consultants in Communicable Disease Control, Environmental Health Officers and others. Outbreaks are then followed up using a standard request form that documents basic details. Participation in the surveillance scheme is entirely voluntary.

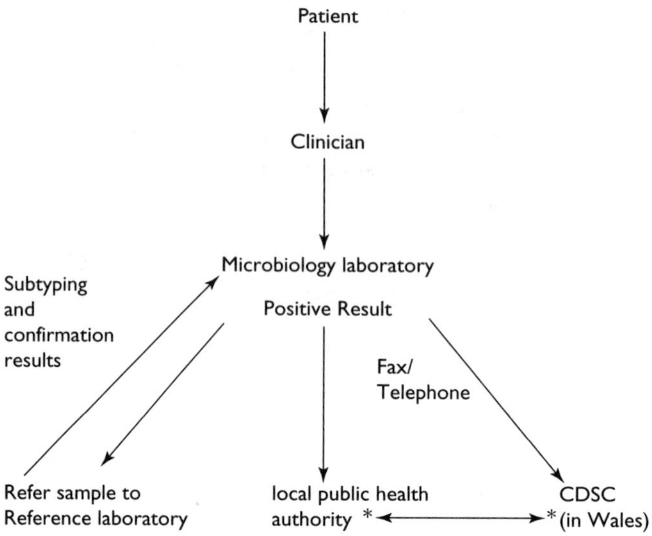

Figure 3.2 Surveillance of VTEC O157

International surveillance

An international surveillance network (Enternet) for human gastro-intestinal infections involving fifteen countries of the European Union, Switzerland, Norway and the United States has been set up. The aim of the network is to conduct international surveillance of human salmonellosis and vero cytotoxin producing *E.coli* O157. This project began first of all with salmonellas in 1994 and was funded from the European Union. The network consisted of the microbiologists in charge of the national reference laboratories in participating countries and epidemiologists responsible for national surveillance. The minimum dataset was collected for each isolate and reported to a central unit. As a result of this network, international outbreaks were recognised. Individual countries identifying apparently isolated incidents were linked together to identify wider outbreaks. One of these outbreaks was twelve cases of O157 in Denmark, England and Wales, Finland and Sweden in visitors to Fuerterventura in the Canary Islands.

HUS surveillance

The UK collaborative study of childhood haemolytic uraemic syndrome began at the end of February 1997 and is conducted by the British Paediatric Surveillance Unit, the Public Health Laboratory Service, SCIEH and the Scottish O157 Reference Laboratory. Paediatricians report cases of clinically diagnosed HUS directly to the PHLS, CDSC or SCIEH by telephone at the time of the diagnosis and also follow this up with a standard report card to the BPSU. Clinical and epidemiological data are collected directly from paediatricians. Routine faeces and serum samples are sent for standard examination and isolation and subtying of VTEC. During the first 2 years of the study, data was collected from 193 clinically confirmed cases, 5 children were reported to have died. One hundred and eighty-six cases presented with a diarrhoeal illness and in 156 of these they suffered bloody diarrhoea. Stool and/or serum specimens were obtained from 185 cases and of these 165 were shown to have suffered VTEC infection.

Incidence in the UK

In the UK the incidence of laboratory reported cases rapidly increased through the 1980s following the first report in 1982. This was certainly due to more and more laboratories introducing screening tests for VTEC. Throughout the 1990s the annual number of cases in England and Wales fluctuated but there is an overall increasing trend. By 1997 the number of reported cases was over a thousand.

A strong seasonal trend can be seen in the reports with about half of the annual total occurring in the third quarter of the year and the lowest number in the first quarter.

The highest incidence in the UK has consistently been in children aged 1 to 4 years.

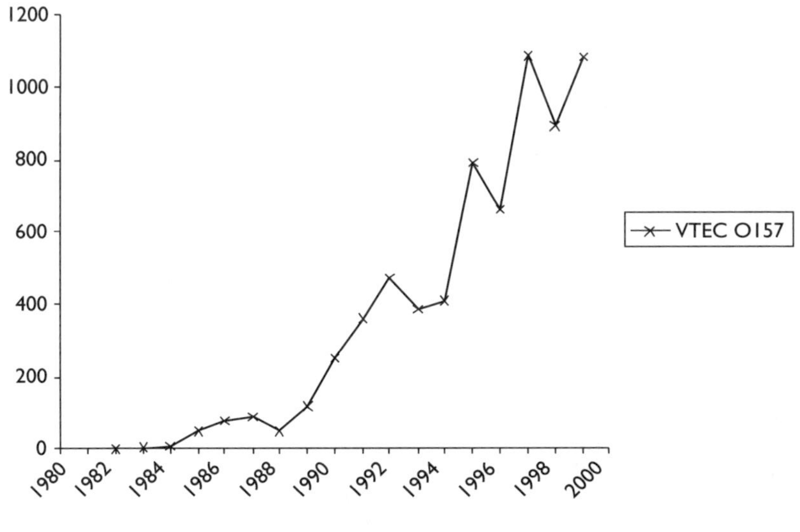

Figure 3.3 Incidence of VTEC O157 in England and Wales, 1982–1999
Source: Public Health Laboratory service data (www.phls.co.uk)

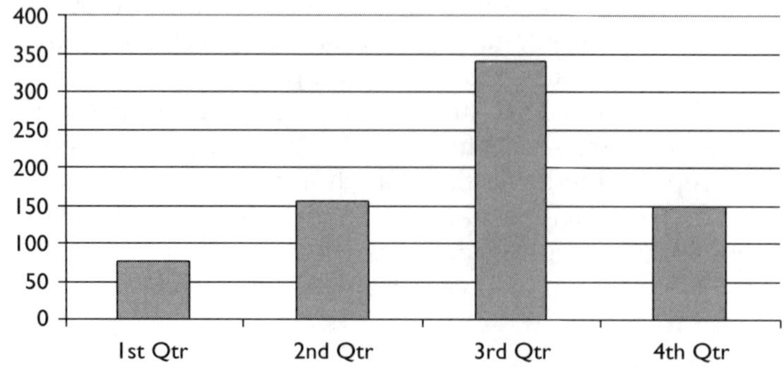

Figure 3.4 Seasonal distribution of cases of VTEC O157 in England and Wales, 1992–1999

Source: Public Health Laboratory service data (www.phls.co.uk)

Outbreaks of VTEC though prominent in the media are relatively uncommon. However, there does appear to be an increasing trend. There is no particularly prominent place where these outbreaks occur. Settings cover private residences, shops, residential institutions, schools, farms and the general community.

The surveillance data have consistently shown that the numbers of reports from Scotland are far higher than those from England and Wales. In part this may be due to its higher profile there. For example, in 1989, *E.coli* O157 was one of a list of thirty-two informally selected organisms which were made 'reportable infections' by laboratories in Scotland, in addition to the statutorily notifiable diseases: this coincided with a doubling (at least) of the recorded incidence in a 12 month period. However, during 1992 and

Table 3.1 Incidence of VTEC O157 in Wales by age group and sex 1990–1998

Age group	*Totals (per 100,000)*	*Males (per 100,000)*	*Females (per 100,000)*
<1	7.9	8.3	7.5
1–4	9.0	10.7	7.2
5–14	1.6	1.8	1.4
15–24	1.4	1.1	1.6
25–34	1.2	1.0	1.3
35–44	0.9	0.8	0.9
45–54	1.0	0.6	1.3
55–64	0.9	0.7	1.1
65+	0.9	0.9	0.8
Total	1.6	1.6	1.5

Chalmers *et al.* (1999).

Table 3.2 *Escherichia coli* O157 general outbreaks England and Wales, by place of outbreak, 1992–1998

Place of outbreak	*92*	*93*	*94*	*95*	*96*	*97*	*98*
Hotel / Guest	0	1	0	0	3	2	4
Hotel / Guest House Residential Pub	1	0	0	2	0	0	0
Restaurant / Café	0	1	0	1	0	2	1
Pub / Bar	0	0	0	0	0	0	0
Mobile Retailer	0	0	0	0	0	0	0
Armed Services Camp	0	0	0	0	0	0	0
Canteen	0	0	0	0	1	0	0
Shop / Retailer	1	2	0	2	2	0	0
Hospital	0	0	0	0	0	1	0
Residential Institution	1	0	0	2	0	1	2
School	1	0	1	1	0	2	1
Community	0	3	2	1	2	2	1
Farm	0	0	2	0	0	4	1
Hall / Caterer	0	1	0	0	0	1	0
Holiday Camp	0	0	0	0	1	0	0
University / College	0	0	0	0	0	0	0
Other	1	0	0	0	1	1	1
Total	5	8	5	9	10	16	11

1993, and against the background of heightened awareness and increased screening, the incidence dropped to about half of the 1991 level (2.3 per 100,000 compared to 4.0 in 1991). The subsequent marked increase, in 1994, also cannot be attributed to changes in reporting rates or screening. This suggests that Scotland does have a higher incidence than the rest of the UK and that there are real significant variations in the incidence of VTEC from year to year (ACMSF 1995: 99; Sharp *et al.* 1994; Waters *et al.* 1994).

To overcome the problem of selective detection of VTEC because of differing policies and methods of different laboratories, a special surveillance study was established in Wales in 1990. All faecal specimens submitted to all PHLS and NHS laboratories have been screened for VTEC using an agreed standard procedure. Positive isolates are reported to the PHLS, Communicable Disease Surveillance Centre (Welsh Unit) in Cardiff.

Between 1990 and 1994 the incidence in Wales remained fairly constant. There was a big increase in 1995 due to one outbreak in a nursery in North Wales with 49 cases. From 1996 to 1998 there was a suggestion of an increasing trend. The highest incidence was in the rural parts of west and north Wales with rates nearly double those in the industrial south-east (Chalmers *et al.* 1999).

Looking further afield, following the recognition of VTEC O157 as a pathogen, the number of outbreaks reported in the USA increased each

Table 3.3 Occurrence and annual incidence of VTEC O157 in Wales, 1990–1998

Year	*Total no. of cases (rate per 100,000 population)*		*No. of sporadic cases (no. of index cases)*		*No. of outbreaks*	*Outbreak summary (setting, mode of spread, 'phage and verotoxin types)*
1990	32	(1.1)	28	(24)	4	Psychogeriatric ward, person-to-person, PT14, VT1&2
1991	39	(1.3)	30	(28)	9	Day nursery, person-to-person, PT49, VT2
1992	41	(1.4)	41	(33)	0	–
1993	34	(1.2)	26	(25)	8	Community, Meat from 1 shop, PT49, VT2
1994	29	(1.0)	29	(24)	0	–
1995	82	(2.8)	33	(30)	49	Day nursery, person-to-person, PT2, VT2
1996	38	(1.3)	38	(31)	0	–
1997	55	(1.9)	51	(41)	3	Home for the elderly mentally infirm, person-to-person, PT2, VT2
					1	Part of a European outbreak
1998	65	(2.2)	65	(47)	0	

year. It was not possible to determine whether this was a genuine increase in incidence, or whether it was a reflection of increased awareness and laboratory screening (Griffin 1995). Preliminary surveillance during 1982 to 1984 had identified 103 cases of haemorrhagic colitis and 28 isolates of VTEC O157 (Remis *et al.* 1984). Studies of laboratories in the USA where all stools were cultured for VTEC O157, recorded that, overall, the organism was isolated from 0.4 per cent of all specimens. In 1994, a nationwide passive laboratory based surveillance system for VTEC O157 was established, through the Public Health Laboratory Information System (PHIS), to the Centre for Disease Control. By 1995, an estimated 29 per cent of clinical laboratories were culturing all diarrhoeal specimens for *E.coli* O157, and the proportion was increasing rapidly. Heightened surveillance, using routine laboratory screening, has led to increased recognition of outbreaks in the USA (CDC 1995a).

In some provinces of Canada, VTEC O157 infection had been notifiable since its discovery as a pathogen. In 1989, haemorrhagic colitis was made reportable to the Laboratory Centre for Disease Control in Ottawa. Overall, an incidence rate of 5.2 per 100,000 was recorded in 1987, 8.8 per 100,000 in 1989, 5.6 in 1991 and 4.06 in 1993. From 1982 to 1989, the number of recorded isolates each year in Canada increased, which could have been due to increased screening. Clusters of HUS were noted in Alberta as early as 1980, and the number of HUS cases has not increased greatly since then, suggesting that the increase in VTEC isolates is an artefact of ascertainment. However, in 1990 and 1991 the numbers decreased, which coincided with control measures publicising correct food handling procedures. These control measures ceased by 1992 and in 1993 the incidence rose again. After a resumption of control measures in 1993, the number of VTEC O157 isolates fell. However, reports suggest that the summer weather was poor in 1993 in Canada, and a decrease in outdoor cooking may also have contributed to the fall in incidence (Waters *et al.* 1994).

Sporadic VTEC O157 infection appears to be more common in Canada than in the USA. In the USA, isolation rates are greater in the northern and western states than in the southern states. A geographical variation is observed in Canada, with Alberta having the highest incidence of 15 per 100,000 in 1989. Not all laboratories in Canada screen for VTEC O157 but systematic screening is carried out by the majority of laboratories in Alberta. This may explain why the highest incidence (16.3/100,000 population) was recorded here. Also, most recorded cases are sporadic, or in family groups. This suggests that the outbreaks seen in other regions represent a small proportion of the total cases, with many cases remaining unrecognised. Nevertheless, the lack of outbreaks in Alberta is difficult to explain (Waters *et al.* 1994).

The age and sex specific incidence rates for Wales, between 1990 and 1993 are presented in Table 3.1. Here, the higher incidence in children cannot be attributed to the selective screening of children's stools. However, it is still likely that younger children with milder diarrhoea submitted stool specimens whilst older children and adults did not. This had the effect of biasing the case series towards children under 5. The bias can be overcome by considering isolation rates where the denominator is the number of specimens tested. This revealed that, whilst isolation rates for children between 1 and 4 years were fairly high, 15 to 24-year-old males and 55 to 64-year-old females also had high isolation rates. This pattern of isolation is largely unexplained.

Detecting outbreaks

Outbreaks of VTEC O157 may be detected by one or a combination of the following routes:

1. Surveillance of the incidence of VTEC O157 (based on laboratory reports) may detect a clustering of cases by time, place or person or by subtype (e.g. an unusual 'phage-type) of the organism.
2. Investigation of sporadic cases(s) may identify further related cases which subsequently are found to have a common source of infection.
3. An outbreak of diarrhoeal illness may be reported directly to the Environmental Health Department/CDC following an event (e.g. wedding/conference/party) and investigators may show that the causative organism is VTEC O157.

Conclusions

Surveillance data underestimated the true incidence of VTEC O157 as patients with mild illnesses did not submit stool specimens and were not recognised as cases. Furthermore, where laboratories selectively screen stools even more cases were missed. Nevertheless, the highest incidence was recorded in Canada and in Scotland, UK. Whilst most surveillance data have indicated an increase in infection since the mid-1980s, these data were considerably biased by improving laboratory techniques and more comprehensive screening policies over the period. In Wales, where all stools were screened between 1990 and 1993, the incidence was fairly constant. There was some indication of geographic variations in Canada, USA and the UK which could not be entirely dismissed as artefactual. Similarly, children aged 1 to 4 appeared to be at increased risk. There was a clear seasonal pattern to infection with the majority of cases occurring during the summer months. Finally, whilst most infection was sporadic, outbreaks were reported that had a major distorting effect on the incidence by time and location.

In England, the Public Health Laboratory of Enteric Pathogens has been recording cases, but up to the mid-1990s these data are likely to have considerably underestimated the true incidence. In 1993 only 13 per cent of laboratories in England and Wales were screening all faecal samples for VTEC O157: however, 79 per cent of laboratories were testing specimens selectively. The selection criteria most widely used were bloody diarrhoea, followed by HUS, colitis, age less than 16 years, a recent history of travel abroad and consumption of meat or beefburgers. Data from the Welsh surveillance suggests that more than half of those infected with VTEC O157 do not have frank blood in their faecal samples so that selective testing will miss these cases. Following the introduction in 1995 of screening for all diarrhoeal stools submitted to the Public Health Laboratories, the incidence in England has approached that in Wales.

In Scotland, the incidence of VTEC O157 is four times higher than in England and Wales and this huge increase cannot be explained by better ascertainment of cases or better outbreak detection. The east of Scotland has tended to have the highest incidence.

In the USA there has been a year-on-year increase on the number of outbreaks of VTEC O157, but it is not possible to determine whether this is a genuine increase in incidence or whether it is due to better ascertainment and reporting.

In Canada, haemorrhagic colitis was made reportable in 1989. The incidence of VTEC O157 in 1989 was 8.8 per 100,000, and by 1991 the rate had fallen to 5.6, which coincided with control measures that publicised correct food handling procedures.

In the United Kingdom, looking at the incidence by age, shows that it is highest in children aged 1 to 4 years and lowest in older people. Also in the UK there is a definite seasonal summer increase in incidence which reaches a peak in August. A similar peak is seen in Salmonella infection, which is usually explained in the growth of the organism on food at higher ambient temperature.

Special studies

In order to get a better appreciation of the true incidence of gastrointestinal infections, a national study, the Infectious Intestinal Disease in England study, was commissioned by the Department of Health. This provides the best data to date of the true incidence of enteric pathogens in the country. The objectives of the study were first to estimate the number of and microbiological causes of cases of intestinal infectious disease in people presenting to their GPs and from whom stool specimens are routinely sent for laboratory examination. Second, to compare these data with the data from the national reporting surveillance system. Third, to estimate the prevalence of asymptomatic infection with these agents. The study used the Medical Research Council's general practice research framework which consists of many research general practices throughout the country.

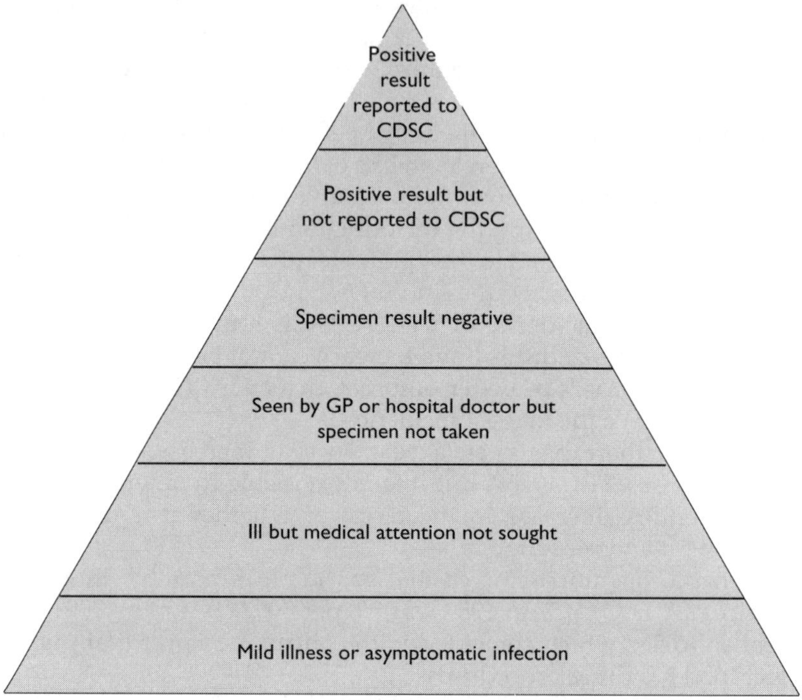

Figure 3.5 Reporting pyramid (A Report of the Study of Infectious Intestinal Disease in England, Food Standards Agency. HMSO. London 2000)

Seventy practices were selected to be representative of the socio-economic characteristics of the area and to represent an urban and rural mix. Cases of infectious intestinal disease were defined as persons with loose stools or significant vomiting lasting less than 2 weeks in the absence of a known non-infectious cause and preceded by a symptom free period of 3 weeks. Vomiting was considered significant if it occurred more than once in the 24 hour period and if it incapacitated the case or was accompanied by other symptoms such as cramps or fever. Non-infectious cases of diarrhoea such as Crohn's disease and ulcerative colitis were excluded. One component of the study was to draw at random a cohort of people registered with the seventy General Practitioners who were followed up for a period of 6 months. These volunteers agreed to fill out diary cards every week and return them to the practitioners. These cards stated whether the person had suffered GI illness or not. If someone developed an illness a stool sample was submitted.

A second component of the study was one where patients identified with a GI illness and controls were selected at random and matched for age and sex within the practice. Unusually in this sort of study controls were asked to provide a normal stool specimen so that the prevalence of pathogens in normal well people could be assessed. Stool samples were dealt with in the most thorough way ever performed in this sort of study. Faecal samples were sent to Leeds Public Health Laboratory and exhaustive range of diagnostic tests were carried out.

The main results of the study were as follows: Salmonellas were identified in 146 (5 per cent) of 2,893 cases identified by GPs. There were 354 Campylobacter cases out of 2,893 (12 per cent). *E.coli* O157 was identified in only 3 out of 2,893 cases (0.1 per cent). Using DNA probes that can identify the vero cytotoxic genes identified 6 cases of non-O157 vero cytotoxigenic, 0.2 per cent out of 27.

Of particular interest in this study was the estimate of the carriage of potential pathogens in otherwise healthy people: 0.4 per cent (10/2,264) controls were found to be excreting Salmonella, 0.7 per cent (16/2,264) were excreting Campylobacter, but none out of 2,264 were excreting *E.coli* O157. These data confirm that VTEC is a rare infection in this country at the moment. From the study the best estimate of the proportion of cases of GI disease caused by VTEC is 0.2 per cent, but in national laboratory reports VTEC comprises 0.6 per cent of laboratory isolates identified by CDSC suggesting that national surveillance overestimates the relevant contribution of VTEC.

Paradoxically the relative rarity of VTEC, together with its clinical seriousness, means that extra attention needs to be paid to the features of the disease so that Environmental Health Officers can recognise it and take appropriate action early. The rarity of the disease means that any single EHO is unlikely to accumulate a significant individual experience of dealing with this dangerous pathogen.

The IID study was able to quantify the reporting pyramid referred to above using the cohort component of the study. For all IID it was estimated that for every case reported to CDSC, 6 patients were investigated by routine laboratory tests, 23 presented with GI symptoms to their GP and there were 136 actual cases in the community. For Salmonellas however, the ratios were much smaller. For every case reported to CDSC it was estimated that there were 2.3 cases presenting to their GP and 3.2 actual cases in the community. Such data could not be provided for VTEC because of the rarity of the organism in the study. The best estimate for the risk of getting a VTEC infection based on this study is 0.03 cases per 1,000/person/years exposed. This compares with 1.57 cases of Salmonella infection per 1,000/person/years exposed and an overall rate of all IID of 33 cases per 1,000/person/years exposed.

A study in The Netherlands from 1996 to 1999, looked at all patients with gastroenteritis who visited their practitioner and compared them with symptomless control patients in a case control study. The study was conducted in 44 practices and covered 1 per cent of the Dutch population. Stool samples were tested for a range of pathogens including VTEC. VTEC was tested by preliminary chain reaction for genes vero cytotoxin 1 and vero cytotoxin 2 and EAE gene and by culture on Sorbitol MacConkey Agar. In the 3 years of the study there were 2,264 cases at a rate of 58 per 10,000 person/years. A subset of these patients were included in the case control study. Of 857 patients providing stool samples, VTEC was identified in 4 (0.5 per cent), and 3 out of 574 controls (0.6 per cent) were also positive. Serotyping showed that only one O157 strain had been identified and that was in a case. Salmonella species were identified in 3.9 per cent of cases and 0.2 per cent of controls. Campylobacter species were identified in 10.5 per cent of cases and 0.5 per cent of controls. This study confirms the low incidence of VTEC in Western countries. It has been estimated that in continental Europe the incidence of shiga-toxin producing *E.coli* is less than 1 in 10^5 and the incidence of HUS is 1.9×10^5 (A Report of the Study of Infectious Intestinal Disease in England, Food Standards Agency. HMSO, London, 2000).

Summary

- VTEC occurs throughout the UK, Europe and worldwide and affects all ages, although incidence is highest in young children.
- Population surveillance is essential for control.
- VTEC in humans is seasonal, with peaks in warmer months.
- Incidence is higher in Scotland than in England and Wales.

KEY MESSAGE

- Develop good communication with laboratories and GPs to identify and investigate cases early.

Chapter 4

How does VTEC spread to humans?

Evidence from sporadic cases

Introduction

Most of the cases of VTEC O157 reported in the UK and the USA and Canada are sporadic, yet information on the sources of VTEC O157 has been predominantly obtained from outbreak investigations. Large outbreaks in particular attract much media and public interest and changes to public health policy may follow. These changes may not necessarily reduce the number of sporadic cases where additional risk factors may play an important role. Data on the sources of sporadic infection with VTEC are limited.

Finding the sources of sporadic infections

By their very nature sporadic cases of infectious diseases are difficult to investigate. Food, water and environmental exposure histories may identify several potential sources of infection; but without further information it is impossible to identify the exact cause of infection. Epidemiological studies gather information from a number of cases of a disease in order to identify risk factors for infection.

This type of epidemiological study is termed an analytical study and the various types of analytical study are described in Table 4.1.

The most appropriate epidemiological method to determine risk factors for sporadic VTEC infection is the case control method. This is because:

1. VTEC O157 is a rare disease and case control studies are particularly suited to the study of rare diseases.
2. Case control studies require few subjects so the studies can be limited to discrete geographical areas, e.g. Wales, England, which is more practical and cheaper than using a wider geographic area.
3. Case control studies can examine multiple exposures of interest in one study.
4. Case control studies are relatively quick to organise and conduct; information on the sources of VTEC O157 is urgently required.

Table 4.1 Observational studies

Type of study	*Description*
Descriptive studies	A description of the health status of a community.
Analytical studies	Hypothesis testing.
Ecological	A study where the units of analysis are populations or groups of people rather than individuals. The study usually relies on exposure data collected for other purposes; exposure and outcome data are compared for different populations. The individual link between exposure and effect cannot be made.
Cross-sectional	A study where the measurement of exposure and effect in individuals is made at the same time. These are useful for investigating exposures which are fixed characteristics of individuals, e.g. ethnicity and blood group.
Cohort	These studies begin with a group of people (a cohort) free of disease, who are classified into sub-groups according to exposure to a potential cause of disease. Variables of interest are specified and measured and the whole cohort is followed to see how the subsequent development of disease differs between the sub-groups. Cohort studies work from cause to effect.
Case control	These studies begin with a group of people with the disease (cases). The proportion of the cases exposed to potential sources of disease are recorded. A second group of people without the disease are identified (controls), and the proportion exposed to the same potential sources recorded. The two proportions are compared.

Adapted from Beaglehole *et al.*, 1993: 36–40; Schlesselman and Stolley, 1982: 8.

5 Case control studies are relatively cheap to conduct.
6 VTEC O157 is serious disease and the case control method presents no risk to the subjects involved.

However, the following disadvantages have to be borne in mind when interpreting the results of case control studies:

- relies on recall or records for information on past exposures
- validation of information is difficult or impossible
- control of extraneous variables may be incomplete
- selection of an appropriate comparison group may be difficult
- rates of disease in exposed and unexposed groups cannot be determined
- method relatively unfamiliar to medical community and difficult to explain
- detailed study of mechanism is rarely possible.

(Schlesselman and Stolley 1982: 18)

Supporting environmental investigations may also be carried out as part of the case control study to identify critical control points for preventing infection.

Analysis and interpretation of case control studies

To interpret case control studies the following questions must be considered

Does the study demonstrate an association between the exposure and disease?

Appropriate statistical tests must be applied to compare cases with controls. Exposure can be given a qualitative (exposed or unexposed) value, which is called a dichotomous variable. Alternatively, it can be quantitative (expressed as exposure limits, or as 'exposure-response relationship'), which can be an ordinal or continuous variable.

Dichotomous categorical data are presented in case control studies as a frequency (2 x 2) table.

The odds ratio is the ratio of the odds of the outcome in the first group compared to that in the second group (*ad/bc*). Where the odds ratio is greater than 1, the exposure is said to be positively associated with the disease, i.e. a risk factor. Where the odds ratio is less than 1, the exposure is negatively associated with the disease, i.e. protective. If an exposure has an odds ratio of 5, exposed individuals are five times more likely to develop the disease than those who are not exposed. Furthermore, the 95 per cent Confidence

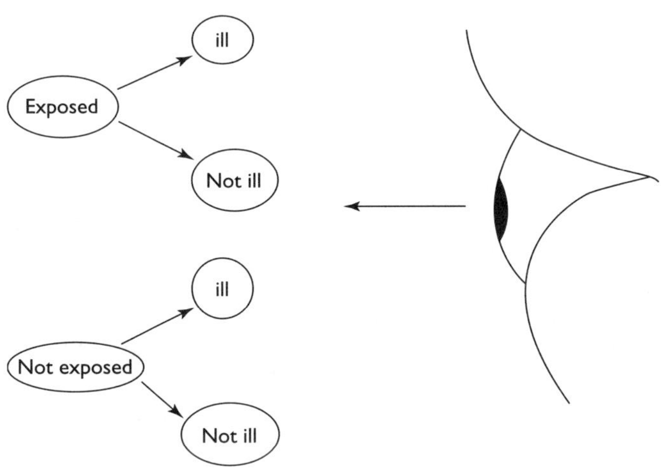

Figure 4.1 Design of a case control study.
Source: Palmer and Galbraith, 1998.

Table 4.2 Contingency table for cohort and case control studies

	Case	*Not case*	*Total*
Exposed	a	b	a+b
Not exposed	c	d	c+d
Total	a+c	b+d	a+b+c+d

Odds ratio = ad/bc
Source: Palmer and Galbraith, 1998.

Interval (CI) can be calculated by a number of different statistical methods. This is presented as a range, e.g. 1.2–5.4 for the odds ratio. There is a 95 per cent probability that the true odds ratio will fall between these numbers.

Could the association have occurred by chance?

The null hypothesis, or H_0, is that there is no association between the disease and the exposure under examination. The probability that the data would have been obtained (by chance) even if the null hypothesis were true is termed the p value. A test statistic (e.g. the Chi square) is calculated. The calculated value of the test statistic is then compared to known distributions (tables of values) which would be expected when the null hypothesis is true to obtain the p value. A cut-off value is chosen for the p value (e.g. 0.05). If the calculated p value is less than the cut-off value, the null hypothesis is rejected. The cut-off value is arbitrary, although by convention it is 0.05. Any value of p below 0.05 is considered statistically significant, and above is considered not significant. Nevertheless, it is not logical to interpret p values immediately above 0.05 as not significant, whilst interpreting those immediately below 0.05 as significant. Therefore, actual p values should be quoted rather than merely stating significant or not significant.

Could the association be due to confounding?

Confounding occurs when the effects of two exposures have not been separated and it is incorrectly concluded that the effect is due to the one rather than the other. Age and sex are common confounding variables. There are several methods to overcome this.

First, a case control study can be designed to take account of confounding variables by matching. Study participants are selected to ensure that potential confounding variables are evenly distributed in the case and control groups. Matching is particularly useful where there is a danger of there being no overlap between cases and controls, e.g. where cases are likely to be younger than controls.

Second, during the analysis of a case control study the data can be stratified into an increasing number of subsets based on exposure to an increasing number of variables. However, multiple cross-classification becomes impossible as the number of variables increases.

In such instances a mathematical model which accounts for the interrelation among variables in the analysis is constructed. The model is a formula with terms representing exposure variables (also called predictor or independent variables) and an outcome variable (the dependent variable). Computer programs are needed to construct these models.

Case control studies are analysed by calculating the odds ratio for each individual risk factor first; this is termed the univariate analysis. Then the odds ratios for risk factors are calculated after taking other risk factors and confounders into account; this is termed multivariate analysis. If a particular variable is found to be a risk factor even after other risk factors and confounders are taken into account it is said to be independently associated with infection.

Could the association be caused by bias?

The biggest threats to the validity of a case control study are: non-participation of subjects and the resulting potential for selection bias; non-differential exposure misclassification typically biasing effect estimates towards the null value; recall bias resulting from differential recollection of past events for cases and controls. The effect of bias in case control studies is considered in more detail below.

The validity of a case control investigation depends partly upon the method by which subjects are selected for study and classified as diseased or non-diseased, and as exposed or unexposed. Bias refers to any systematic error in the design, conduct or analysis of the study that results in a mistaken estimate of an exposure's effect on the risk of disease.

The case control method is susceptible to several sources of bias, most of which can also occur in cohort studies. An outline of the major categories of bias is given below:

Ascertainment and selection bias

Exposures associated with differential surveillance, diagnosis, referral or selection of individuals can lead to biased estimates of relative risk.

The control group is intended to provide an estimate of the exposure rate that would be expected to occur in the cases, if there were no association between the study disease and exposure. Controls must be free from disease and similar to the cases with regard to past potential for exposure during the time period of risk under consideration.

Controls can be selected from hospitals or from the community. Community controls are sometimes selected from the friends, families or neighbours of the case. Alternatively, they are selected from a sampling frame that conceptually is a list of all potentially eligible controls in the population under study. Hospital controls are relatively easy to recruit and may have a similar motivation to cases in recalling past events. However, they are not necessarily representative of the population from which the cases arose. Controls selected through their relationship with the case, e.g. friends, neighbours, etc. will tend to resemble the cases in lifestyle which may conceal the very difference sought by the study.

Probability sampling procedures (e.g. random or systematic sampling) are then used to ensure that every potential control has an equal opportunity of being selected. A sampling frame that does not provide a complete coverage of the target population may be biased and lead to bias of the control group.

Bias in the measurement of exposure

This includes recall bias, where the case is more likely to remember exposure than controls. Interviewer bias may occur where the interviewer, perhaps subconsciously, encourages certain responses to questions posed.

Furthermore, for VTEC O157 (a bacterial acute disease) there is a 'critical time window' (or 'aetiologically relevant exposure period') within which, occurrence of a particular exposure may be relevant to causation of the disease. If exposures outside that time window are included, it can lead to a misclassification of exposure and reduce the likelihood that an existing association will be observed.

Determining exposure to risk factors for VTEC O157, therefore, requires the documentation of past exposure. Methods of exposure measurement in epidemiology are: personal interview, self-administered questionnaire, diaries, observation by an investigator, records, physical/chemical measurements on subjects or the environment.

The logic of the case control study requires that exposure measurements permit inferences about past exposures. Thus, personal interviews, self-administered questionnaires and references to records about subjects have been the methods most commonly used.

Misclassification

The determination of an individual's exposure status or disease may be subject to error. A clear case definition and definition of exposure reduces this source of bias.

Do the non-associations arise as a result of insufficient power?

A case control study should include sufficient subjects (cases and controls) to avoid two sources of error: (1) claiming that an exposure is associated with disease when, in fact, it is not; and (2) claiming that an exposure is not associated with a disease, when, in fact, it is. The first error is referred to as a Type I error or false-positive. The second error is a Type II error or false-negative. The probability of making the first error is called the level of significance and is denoted alpha. The probability of making the second error is denoted beta, and the quantity 1–beta is the power of the study. Assuming that a given risk factor really is associated with disease, the power (1–beta) of a study is the probability of finding that the sample estimate of relative risk (indicated by the odds ratio) differs significantly from unity. Four values are needed to calculate the number of subjects required: (1) the relative frequency of exposure among controls in the study population; (2) a hypothesised relative risk associated with exposure that would have sufficient biologic or public health importance to warrant its detection; (3) the desired level of significance (alpha) and (4) the desired study power (1–beta). Formulae and computer programs are used to perform the calculation. In practice, the size of a study is restricted by resources (financial, manpower, goodwill), the number of available cases and time. The need for prompt answers to a public health problem and a low disease incidence will limit the size of the case series. As a result, the sample size analysis often begins with a fixed number of cases and then determines the resulting power (1–beta) to detect various values of relative risk. Such calculations can also be used to calculate the power of a completed study.

Case control studies of sporadic VTEC O157 infection

The first population based case control study to examine risk factors for food, person-to-person and zoonotic transmission of sporadic VTEC O157 infection in the UK was published in 1998. Cases were defined as any individual submitting an acute initial faecal specimen, from which VTEC O157 was isolated and which were the index case in the household (first case to be reported) and not part of a recognised general outbreak.

Eighty-five cases were compared to 142 controls matched for age group (<1, 1–4, 5–14, 15–24, 25–34, 35–44, 45–54, 55–64, 65+) and sex, randomly selected from the lists of patients at the same GP as the corresponding case.

Relevant exposures (and potential confounders), recorded at interview using a standard questionnaire were food: beefburgers, cooked sliced meats, eating out, untreated water, unpasteurised milk; animals: handling farm animals, handling farm animal faeces, living on a farm, visiting a farm, household member with an occupation involving farm animal (manure) contact; other: attending children's day care, household member with a previous diarrhoeal illness, household contact under 5 years, staying away from home, bathing/paddling in recreational water. Methods of cooking beefburgers at home were recorded. Implicated catering establishments were visited and the policies for ensuring thorough cooking of beefburgers recorded. Sources of ready cooked sliced meats (either retail or catering), were visited, the opportunities for cross-contamination assessed using a standard proforma which was developed for the study. Manure samples or rectal swabs were obtained to look for carriage of VTEC O157 by animals with which the case had been in contact.

Two risk factors for foodborne transmission were identified. Consuming a cooked sliced meat meal from a caterers, such as a cooked sliced meat sandwich/roll was associated with infection in the univariate analysis. Consuming any beefburger and, in particular, consuming beefburgers cooked at home was statistically protective. Beefburgers cooked in catering premises other than one particular fast food chain (Fast Food Chain A) were associated with an increased risk of infection, suggesting that Fast Food Chain A adopts different, and more effective, procedures for ensuring adequate cooking. Insufficient cases and controls had eaten at any other specific restaurant chains to allow any conclusions similar to those for Fast Food Chain A. A comparison of the practices observed to ensure thorough cooking in Fast Food Chain A when compared with the other caterers would support the finding of elevated risk associated with premises other than those of Chain A. Thus the critical control point for beefburger preparation is thorough cooking.

Two farm/farm animal exposures were independently associated with an increased risk of infection. Visiting a private farm as opposed to an open farm was independently associated with infection. Also, VTEC O157 was isolated from cattle on two farms connected with human cases. The strains were indistinguishable from the human isolates. Furthermore, having a household member whose occupation involves farm animal contact was associated with infection independently of the other farm exposure variables. Hence, the household contact is likely to be bringing contamination into the home rather than taking the case to the farm with them.

Contact with a household member with diarrhoea was independently associated with infection. What the various risk factors identified for sporadic cases have in common is that they operate near the cases. This implies that, irrespective of more sophisticated interventions, local inspection directed at catering establishments serving cooked meats and at burger restaurants and public education to prevent spread on farms and in houses would reduce the burden of disease (Parry *et al.* 1998).

A later case control study conducted in England between 1996 and 1997 compared 369 cases with 511 controls. Similar to the Welsh study, the most striking risk factors were contact with people with diarrhoea, visits to farms, travel within the UK, eating out and exposure to recreational water (Adak *et al.* 2000).

A further UK study conducted in Scotland between 1996 and 1999 also identified visits to farms and contact with animal faeces as independent significant risk factors for sporadic VTEC O157 infection (Locking 2000).

Earlier studies conducted outside the UK have identified different risk factors. In Canada, a case control study was conducted in five cities between June and September 1990. The 110 cases were compared to 220 neighbourhood controls matched for age (by 5 years) and sex. Consuming ground beef that was pink was the most important risk factor (Odds Ratio (OR) = 3.5; 95 per cent Confidence Interval (CI) = 1.8–6.9) although attending a picnic or special event away from home was also significantly associated with illness (OR = 1.9; 95 per cent CI = 1.1–3.3). In contrast, eating ground beef in casseroles was statistically protective (OR = 0.5; 95 per cent CI = 0.3–0.8). It is likely that individuals who ate casseroled meat (which was generally cooked twice and not pink in the middle) were less likely to have eaten the other types of ground beef. An earlier Canadian case control study (1986–7) compared 81 cases to 96 controls who had diarrhoea but whose stool samples did not yield VTEC O157 (age matched). In a separate analysis 52 of the cases were compared to 49 neighbourhood controls. In the first analysis cases were more likely than controls to have eaten hot dogs but were not more likely to have eaten them frequently. In the second analysis, cases were more likely to have eaten barbecued food ($p<0.10$) and undercooked meat ($p<0.05$) (Bryant *et al.* 1989). A further case control study conducted in New Jersey in 1994 identified consuming hamburgers in the 7 days prior to onset (MOR = undefined; lower 95 per cent CI = 3.3; $p<0.001$). Also, 80 per cent of the 35 hamburgers eaten by cases had been prepared at home or at private picnics. Food preparers in case households were more likely than controls to report not washing their hands after handling ground beef (MOR = 8.5; $p<0.005$) and not washing work surfaces that had been in contact with raw ground beef (MOR = 10; $p<0.05$). Food preparers in case households were also more likely to report placing cooked hamburgers back onto an unwashed plate which had been previously used for raw ground beef, but this association was not statistically significant. The evidence of poorer food handling in the homes of cases together with the finding that eating undercooked hamburgers was not associated with infection indicates that the hamburgers had cross-contaminated other foods and that the hamburger itself was not necessarily the vehicle of infection (Mead *et al.* 1997).

Microbiological investigations of sporadic cases

The investigation of some sporadic cases of VTEC O157 has included obtaining environmental and veterinary samples for microbiological analysis. In Scotland and Canada, identical strains have been isolated from patients and the implicated cattle (Anonymous 1994; Coia *et al.* 1995;

Renwick *et al.* 1993; Synge and Hopkins 1992). Martin *et al.* (1986) isolated VTEC O157 from a healthy heifer on a farm where two cases of HUS had drunk raw milk. Other studies have confirmed the role of zoonotic transmission of VTEC O157 in the UK, and in particular, the role of companion animals, such as dogs, in carrying the infection to humans who might otherwise not have been exposed. Trevena *et al.* (1996) report the isolation of indistinguishable strains from a child (male, 1 year), a pony and dog at a smallholding, which he had visited. The child had touched only the dog. The investigators suggest that the close relationship between farm animals, companion animals and humans on smallholdings may increase the risk of transmission (Trevena *et al.* 1996). Direct transmission from animals to humans is now reported with greater certainty as PFGE can confirm identical strains in the human case and implicated animals (Rice *et al.* 1996).

Investigating sporadic cases of VTEC O157

Cases must be investigated urgently

A sporadic case of VTEC O157 is most likely to be notified to an Environmental Health Department by a microbiology laboratory. This is preceded by the case visiting their GP or being admitted to hospital and a faecal sample being submitted and analysed. A significant delay of several days, possibly weeks, is inherent in the reporting system. Therefore, notified cases should be investigated urgently to prevent any further delays and associated continuing risks to public health.

Case reports should be received by telephone or fax (see Chapter 2) and the investigation started as soon as they are notified.

Aims of sporadic case investigations

Environmental health investigation of sporadic VTEC O157 infection should achieve four aims. First, information on potential sources of infection is gathered to determine whether a continuing source of infection exists which presents a risk of infection to other individuals. Second, the likelihood of the apparently sporadic case being part of an outbreak is estimated by using the information on potential sources as well as further information on other suspected cases. Third, action may need to be taken to prevent secondary spread from the index case by preventing them attending an environment where further spread is likely, for example, preventing a child attending pre-school day care. Finally, advice and guidance on preventing secondary spread in the home environment should be given including advice on personal and food hygiene. It is also likely that the case will want to ask questions about the organism and how it is spread.

This opportunity can be taken to provide general information on food hygiene and on hygiene during pet and animal contact.

Gathering information on potential sources

A food/water consumption history and activity history for the 1 to 7 days preceding onset of symptoms should be carefully obtained and recorded. It may also be useful to ask about activities up to 14 days prior to onset as some experts believe that the incubation period for VTEC O157 can be up to 14 days. In many instances this will require the case or their carers to recall events which occurred several weeks before and inevitably they will not find this easy. The following prompts and techniques can be used to help them remember.

1 Work from the most recent date (i.e. the day before onset) backwards.
2 Encourage the case to use diaries or calendars during the interview to aid their memory.
3 Does the incubation period include significant events such as school holidays, bank holidays, periods when the weather was particularly warm (which might have coincided with a family day out/picnic/BBQ, etc.) or cold, windy, and so on.
4 Does the case work shift patterns and if so what were their shifts at that time? This frequently affects their eating patterns.
5 If the case is happy to consult other members of their family or friends encourage them to do so. Contact them again the following day to see if this has uncovered any further information.
6 Leave a food/activity history with them for them to think about. Also leave your telephone number so that they can contact you if they remember anything significant.
7 Small children in particular tend to have fairly constant diets so make a record of foods which the child regularly eats every day or week as they are very likely to have eaten these during the incubation period.
8 What is their job and that of other household members and do they bring them into contact with any of the potential sources listed below.
9 What are their hobbies and do they bring them into contact with any of the potential sources listed below.

Record carefully where and when any meals out, including takeaways, were purchased. Local knowledge is very helpful in identifying food premises. The retail source of foods eaten at home should be recorded. The following information on exposure to potential sources should always be recorded:

- Details of meals eaten and water consumed (is the domestic supply untreated and have there been any water supply problems?). Note any unusual meals such as picnics or BBQs.
- Details of contact with animals (pets and other animals) in and outside the home.
- Contact with a person with diarrhoea or human faeces.
- Contact with animal (including pet) manure (gardening, farming, walks in the country).
- Foreign travel.
- Travel in the UK.
- Recreational water use (did they immerse head/swallow water).

Particularily ask about anything new or different, e.g. new pet, new job, new hobby, new activity.

Further investigations

Follow-up of information gathered during sporadic food poisoning investigations varies widely between local authorities and between individual officers. Obviously, each case must be judged on its merits.

The general principle is that if the officer suspects that a continuing source of infection exists they should investigate further.

For environmental sources of infection, e.g. recreational water, farm or other animals, this would involve visits to the premises concerned and an assessment of the risk management procedures in place (see Chapter 7 on Control).

If environmental samples (e.g. water samples or animal faecal samples) are required, it is important to discuss this with the microbiology laboratory prior to sampling. Special techniques can be employed to increase the likelihood of isolating VTEC O157 from such samples and samples may need to be collected in a particular way.

The local Veterinary Laboratory Agency (see list of useful contacts at end of book) may also be able to assist in investigating particular herds of animals.

In all situations it is up to the EHO to make the case, if they believe it is necessary, for further investigations based on the need to protect public health.

For food sources, risk assessments of food handling procedures at commercial establishments may be appropriate. Several critical control points for preventing VTEC O157 have been identified (see Chapter 7). For example, the case control study detailed above clearly showed that adequate cooking is a critical control point in preventing viable VTEC O157 remaining in beefburgers. Risk assessments in food premises following sporadic VTEC O157 investigations should not only record usual procedures for ensuring the adequate cooking of burgers but also any unusual events, such as equipment breakdown, etc. during the implicated time period.

Food samples may be obtained at the home of the case or from commercial establishments. Very rarely will food from the same batch as that consumed by the case still be available for examination. Again, it will be necessary to make a case to the microbiology laboratory that the food should be examined. Particularly sensitive techniques will need to be employed as the numbers of organisms in any food sample are likely to be low (see Chapter 6).

Is the sporadic case part of an outbreak?

Any meals or outings which the case shared with other individuals are a potential source of infection and the case may be part of a wider outbreak. Outbreaks are identified at a number of levels. At the national level, the Public Health Laboratory Service (PHLS) Laboratory of Enteric Pathogens and Communicable Disease Surveillance Centre (CDSC) maintain national surveillance of VTEC O157 and will spot any increase in cases by region as well as increases in particular strains of the organism (see Chapter 3). However, this can only be achieved retrospectively, once laboratory results are available. At the local level, EHOs investigating sporadic cases are ideally placed to identify outbreaks very quickly even before all the laboratory reports have been received. The search for further cases should begin with the patient's immediate contacts and those outside of their immediate family who shared meals out and activities (e.g. visits to farms, etc.). If the case attended a festival or event organised outside the local authority boundary, it may be necessary to liaise with colleagues in other regions to determine whether they are investigating cases associated with the same event. This can require just a simple telephone call but can mean the difference between an outbreak being promptly identified and controlled rather than a source of infection continuing to endanger public health.

Finally, VTEC O157 is a rare disease and each local authority can expect to investigate only a small number of sporadic cases per year. If several sporadic cases are reported together, a search for a common source of infection should be rigorously pursued. CDSC can assist in determining whether the cases are grouped by time (onsets close together), place (home addresses or activities) or person (similar age groups). Chapter 5 covers the investigation of outbreaks in detail. When outbreaks are identified, clear and detailed sporadic case histories are invaluable in the early stages of outbreak investigation and control.

Preventing secondary spread

Very few organisms are needed for VTEC O157 to cause symptoms in humans. Therefore, secondary spread is a real possibility and far more likely than for cases of Salmonella and Campylobacter (see Chapter 1). Details of the cases occupation(s) or, for children, any school or day-care arrangements should be recorded. Part-time and informal arrangements should be included. Detailed national advice is available on which categories of persons should be excluded from work/school and for how long. (Subcommittee of the PHLS Advisory Committee on Gastrointestinal Infections, Guidelines for the control of infection with vero cytotoxin producing *Escherichia coli* (VTEC) Communicable Diseases and Public Health 3:14–23).

It may also be useful to discuss the particular situation with the local CCDC or CDSC. If a child is excluded from day care it is important to ensure that any alternative arrangements for their care do not also present a risk of further spread. For adults, all occupations, including part-time and voluntary work, must be taken into account.

It may be appropriate to screen family contacts for VTEC O157 particularly if they are symptomatic and/or are in a situation where they could spread the disease further.

Some research has been carried out to determine the chances of household spread of VTEC O157. No secondary cases occurred where the index case was an adult. The most common pattern of transmission observed within the household is from an index case under 1 year to household contacts aged 1 to 4 and adults aged 15 to 24. Overall, household contacts aged between 1 and 4 years and adults between 15 and 34 are most likely to contract VTEC O157 from an index case. Children between 1 and 4 years old explore the domestic environment and frequently put their hands (and other objects) in their mouths. Adults aged between 15 and 34 are likely to be looking after children suffering from diarrhoea. When contacts who are actually parents of an infected child were compared to the remainder of contacts an elevated risk of contracting VTEC O157 in the home was observed. The pattern of transmission suggests person-to-person spread via the faecal–oral route rather than foodborne transmission. Children under 1 year are most likely to transmit VTEC O157 to other household members followed by children aged 1 to 4 and children aged 5 to 14. Furthermore, the decreasing transmission rate with the age of the child reflects the decreasing level of help with toileting that a child requires. A matrix of household transmission of VTEC O157 is given in Table 4.3.

Advice on preventing secondary spread and future infection

Providing practical advice on how to prevent the spread of VTEC O157 in the home is an essential part of the sporadic case investigation particularly

Table 4.3 Matrix of household transmission of VTEC O157

	Age group of index case (years)				
Age group of household contact (years)	*< 1*	*1–4*	*5–14*	*15+*	*Totals*
<1	0/0 0	0/3 0	0/1 0	0/0 0	0/4 0
1–4	1/2 50%	3/8 38%	1/3 33%	0/3 0	5/16 31%
5–14	0/0 0	1/11 9%	0/7 0	0/8 0	1/26 4%
15+	2/6 33%	5/49 10%	2/19 10%	0/61 0	9/135 7%
Totals	3/8 40%	9/71 13%	3/30 10%	0/72 0	15/181 8%

Source: Parry and Salmon, 1998.

if the case is still experiencing symptoms. This will mean explaining in plain terms how the organism can spread via a faecal–oral route to other family members possibly via surfaces in the bathroom, etc. The advice should be tailored to the particular circumstances of the family, but the following principles will usually apply. When a patient with loose stools uses a toilet an aerosol is created which can contaminate the immediate area (toilet bowl, seat, etc.). Also, any surface which they touch before they can wash their hands will become contaminated. Therefore the following precautions are recommended:

ADULTS AND OLDER CHILDREN

- If there are two toilets in the household, the most effective means of preventing spread is for the case to use one toilet and for everyone else to use the other.
- Where this is not possible, the toilet seat, cistern and flush handle should be wiped after use with the appropriate dilution of a commercial disinfectant/bleach.
- The case should then wash their hands thoroughly preferably using separate soap and towels to the rest of the family.
- The case should wash their hands thoroughly using soap before preparing food.

CHILDREN/ADULTS REQUIRING HELP WITH TOILETING

- Younger children and adults who need help with toileting present a particular risk of infection to their carers. The principles outlined above for adults should be followed with the carer also washing their hands thoroughly.

YOUNGER CHILDREN IN NAPPIES

- Soiled nappies and linen present a risk of infection. Nappy changing should preferably take place on a washable surface which can be wiped down afterwards with an antibacterial solution which is safe for surfaces which children are in contact with.
- Disposable nappies should be sealed carefully in leak-proof bags. Cloth nappies should be washed as soon as possible, preferably in an automatic washing machine.
- Hands should be thoroughly washed, using soap, after nappy changing.

General advice on preventing future infections

It is extremely unlikely, although not impossible, that an individual will get a repeat infection of VTEC O157. Nevertheless, cases and their carers are generally very anxious to know how to prevent further infections. The EHO is the best source for this information. Currently recognised routes of transmission from a predominantly bovine reservoir to humans are summarised in Figure 7.1 (p. 86). Further details are given in Chapter 5 and in Chapter 7. General printed information on VTEC O157 is available from the PHLS (CDSC) and local groups may consider producing leaflets.

KEY MESSAGES

- Routes of sporadic infection of VTEC O157 are mainly person-to-person and zoonotic spread. Foodborne transmission is mainly via foods eaten outside the home. Cooking beefburgers adequately is a critical control point in catering premises.

CHECKLIST FOR INVESTIGATING SPORADIC INFECTIONS

For the 7 days prior to onset record

- Detailed food/water consumption with meals out highlighted and retail sources recorded.
- Contact with animals/pets and their manure by case or other family members.
- Travel abroad and within the UK.
- Contact with people with diarrhoea or human faeces.

Also record:

- Could others have been exposed? Do you need to visit any commercial premises/take samples/liaise with other agencies?
- Are any other cases suspected? Investigate immediately.
- Advise on secondary spread in the home.
- General personal and food hygiene advice.

Chapter 5

How does VTEC spread to humans?

Evidence from outbreaks

Investigation of outbreaks provide an important opportunity to identify vehicles of infection by VTEC, and to clarify its epidemiology. Thorough systematic investigation is now an essential requirement of public protection, but in addition every opportunity should be taken to exploit the 'natural experiment' that outbreaks afford to learn new knowledge.

Food and drinking water outbreaks – beef products

The first two community outbreaks of VTEC O157 in the USA were associated epidemiologically (via a case control study) with beefburger patties from a single chain of fast food restaurants, in Oregon and Michigan in 1982. There was an indication that the grill at one of the restaurants did not reach the specified temperature (as stated by the fast food chain) during busy periods. Whilst food samples taken from the implicated restaurants were negative, subsequent examination of beefburger meat, retained at a processing plant for quality control purposes, yielded *E.coli* O157. It is thought that a failure to use the patties in lots accounted for a prolonged period of exposure. Despite case finding efforts in the community, an attack rate of only one case per 1,000 burgers was estimated (Riley *et al.* 1983; Wells *et al.* 1983).

Subsequent community outbreaks, in the USA and Canada, have also been associated with beefburgers. Ground beef was responsible for 22 (58 per cent) outbreaks of VTEC O157 in the USA between 1982 and 1994 and VTEC O157 was isolated from the implicated meat in six outbreaks. State-wide outbreaks have been recognised, with contaminated lots of meat being dispersed over vast geographical areas. The recognition of such community outbreaks, in the absence of awareness of, and routine screening for, VTEC O157, would have been extremely difficult, and it is likely that other similar outbreaks have been missed in previous years.

The largest outbreak recorded to date in the USA occurred from November 1992 to February 1993 in four states: Washington, Idaho,

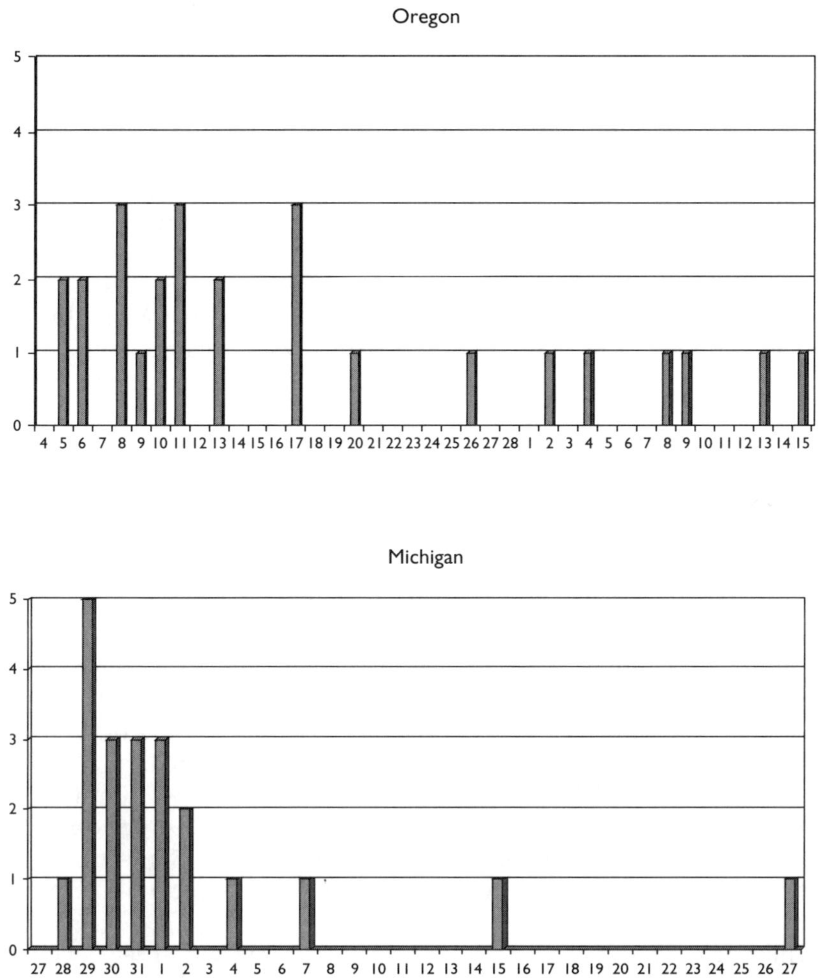

Figure 5.1 Cases of haemorrhagic colitis, according to date of onset in first recognised VTEC outbreaks, USA, 1982

Source: Riley *et al. The New England Journal of Medicine*, 1983: 308; 12: 681–5.

California and Nevada. More than 500 laboratory confirmed cases were recorded, and four associated deaths occurred. Three separate case control studies, in three states, all implicated beefburgers from a single chain of fast food restaurants. The outbreak strain was also implicated from eleven lots of patties, which had been distributed to restaurants in all the affected states. Meat from five slaughter plants in the USA, and one in Canada, provided meat for the suspected batch. Early recognition of the outbreak in Washington (where there is heightened awareness, due to the previously described outbreak, and where VTEC O157 infection is a reportable illness) meant that a multi-state recall of unused beefburger patties was initiated (CDC 1993). A further outbreak was associated with beefburgers, which were served at a country club. Again, inadequate cooking was identified, and the outbreak strain was isolated from remaining frozen raw hamburger patties (Roberts *et al.* 1995).

Beefburger associated community outbreaks have not been confined to restaurant-served beefburgers. A small outbreak in California in 1993 was traced to a supply of beefburgers from a local supermarket chain, where large tubes of ground beef were purchased from an outside supplier, to which trim meat from other sources was sometimes added. VTEC O157, of the same 'phage type as the human isolates, was isolated from leftover ground beef at the home of one of the patients, and from ground beef which was available for sale at the supermarket. VTEC O157 of a different 'phage type was isolated from ground beef at a further implicated supermarket. At least one of the affected families had eaten the beefburgers 'medium rare'. The local health department subsequently issued a press release advising consumers to cook ground beef thoroughly (CDC 1994).

In the UK, in 1991, an outbreak associated with a fast food chain of restaurants primarily located in the Preston area, was also associated with beefburgers. The outbreak attracted considerable media attention and prompted the Chief Medical Officer in February 1991 to issue advice to the public and to food handlers (via Local Authority EHOs) regarding the adequate cooking of beefburgers (DOH 1991). Foodborne outbreaks in institutions involving minced beef have also been reported.

In the Republic of Ireland, in 1995, an outbreak was identified in a developmental centre for handicapped children. In all, 62 per cent of the 24 children tested had serological evidence of *E.coli* O157 infection. Meals containing ground beef were the suspected vehicle of infection (O'Donohue 1996).

Whilst ground beef has most often been implicated in outbreaks, contaminated rare roast beef has also been the cause of outbreaks.

Food and drinking water outbreaks – milk and dairy products

Several outbreaks due to the consumption of unpasteurised milk have been reported. In the UK, investigation of six confirmed cases of VTEC in the Trent region in May 1993 linked cases with consuming unpasteurised milk from a single farm. VTEC O157 strains, of a similar 'phage type and VT type, and all harbouring a 92 kb plasmid, were isolated from the human cases, from rectal swabs taken from 10 cattle at the farm, and from a sample of milk from one of the positive cows suggesting the farms were the source of infection. The milk may have been contaminated by faecal material or VTEC O157 may have been excreted in the milk (Chapman *et al.* 1993; Wall *et al.* 1996). More recently, three cases of VTEC O157 were associated with the consumption of untreated milk produced locally in Yorkshire. An identical strain to the human isolates was also recovered from a bottle of unpasteurised milk retained by one of the cases (CDSC 1996).

A large community outbreak (seventy cases confirmed by culture or serology) was reported in Scotland in 1994. Epidemiological investigations revealed that 90 per cent of the cases had consumed pasteurised milk from a local dairy. Later, a strain of VTEC O157, identical to those isolated from human cases (PT2, VT2), was isolated from a pipe, which carried milk from the pasteurisation plant to the bottling machine, and a discarded bottling machine rubber. Identical isolates (confirmed by PFGE) were made from a bulk feed tank, containing pooled milk from feeder dairy farms, and from bovine samples from one of the farms. This is the first ever recorded outbreak involving a pasteurised milk supply, although the preliminary report does not indicate whether the contamination occurred as a result of cross-contamination of the final product, or because of an inadequate pasteurisation process. There is no evidence that VTEC O157 can survive normal pasteurisation temperatures (Upton and Coia 1994).

Cheese made from unpasteurised milk was related to at least twenty cases of VTEC O157, PT28, in Grampian in 1994. Grampian Health Board issued a public health warning, relevant cheeses were withdrawn from sale and production stopped. As in the above outbreak, a primary producer (i.e. a farm) was implicated. VTEC O157 was isolated from the cheese (Anonymous 1994; Curnow 1994).

In 1991, an outbreak of sixteen cases in the North West of England was associated in a case control study with locally produced live yoghurt. Cases were spread over a period of about 7 weeks. Sophisticated methods of strain discrimination were not available and, as the outbreak involved one of the commonest 'phage types in the UK (PT49), the early cases in September may have been unrelated. The report indicates that batches of 200 dozen pots of the implicated yoghurt were produced once a week. No shelf life was indicated. It was thought that one batch could have accounted for all the cases, especially as no cases were detected in persons who had consumed other batches of flavoured and unflavoured yoghurt.

The environmental investigation revealed several problems. Opportunities for cross-contamination of pasteurised milk by raw milk were noted and records of the pasteurisation times and temperatures were not kept. Assuming that cross-contamination did occur, there was little opportunity for rapid multiplication as the yoghurt was stored and delivered to the retailer under refrigeration at below 8°C. However, poor temperature control may have been exercised by the *retailers* and/or the *consumers*.

A further possibility is that *E.coli* O157 numbers decrease in yoghurt as is the case with other *E.coli*. Most cases were in children under 10 years of age. This age group are more likely to be recognised as cases as they develop more severe disease (such as HUS) and are more likely to be presented for medical attention. Nevertheless, the implicated yoghurt was marketed specifically for infants and it may be that more children were exposed. This may also account for the relatively high rate of HUS development. Alternatively, despite active case finding, many community cases may have been missed, as they did not present for medical attention. This would also explain why so few cases were identified when 200 dozen pots of yoghurt were produced.

Samples of the yoghurt (different batch), and pasteurised and unpasteurised milk obtained from the manufacturer were tested for VTEC O157 and were negative. Rectal swabs, obtained from the herd which supplied milk for the yoghurt, were also negative for VTEC O157 although other serotypes of VTEC were isolated (Morgan *et al.* 1993).

An outbreak associated with fresh cheese curds was identified in Wisconsin in 1998. Unpasteurised cheese had inadvertently been used to make fresh cheese curds which were subsequently incorrectly labelled as 'pasteurised'.

Food and drinking water outbreaks – other foods

Handling raw vegetables (particularly potatoes) was thought to be the cause of the first recognised community outbreak of VTEC O157 in the UK in 1985, although the source of contamination of the vegetables could not be established (Morgan *et al.* 1998). More recently, a cluster of cases in Maine, USA, was linked to vegetables grown in a manured garden (Cieslak *et al.* 1993).

Fresh pressed apple juice was implicated in an outbreak affecting at least twenty-three people in Massachusetts in 1991. Samples of the cider, and environmental samples at the mill, were negative. The most likely source of the contamination was debris on the ground contaminating the apples, which were not washed thoroughly before being crushed. The mill owner kept cattle at the mill (Besser *et al.* 1993). Interestingly, in 1980, in Canada, before VTEC was discovered, an outbreak of diarrhoea and HUS was also associated with the consumption of fresh apple juice. Apple juice would not have been considered as a likely vehicle of infection in any outbreak, particularly since it is quite acidic. Only by an open minded and thorough investigation could the vehicle be detected (Steele *et al.* 1982).

A single brand of dry-cured salami (manufactured without cooking) from a local grocery chain in Washington State was identified by a case control study as the vehicle. The outbreak strain was isolated from samples of the implicated salami that had been voluntarily withdrawn from sale in that state. Subsequently, the manufacturing company voluntarily withdrew the product, which had also been distributed in Oregon and California. Later, samples of the salami collected in California were also positive. Identical strains were isolated from an individual (in Washington) who had not eaten salami, but who had eaten turkey from the same supermarket delicatessen, indicating cross-contamination. PFGE was invaluable to the epidemiological investigation, differentiating the outbreak strain from others circulating in the community (CDC 1995b).

Cooked sliced meats and cooked meat products (sandwiches/salads) have been implicated in several outbreaks. Cross-contamination is thought the most likely cause of such outbreaks.

In May 1996, a large outbreak of VTEC O157 occurred in the Okayama Prefecture in Japan. In all, 468 patients were affected and there was one death. Similar outbreaks were subsequently detected in three other areas. In 1996 a major outbreak was detected in primary schools in Osaka Prefecture. At least 6,000 school children were affected. It is estimated that, overall, 8,000 people suffered symptoms and (as of July 1996) five have died. VTEC O157 was isolated from raw vegetable salad in one of the outbreaks and from raw beef liver in another. Later reports indicate that fresh radish sprouts were the food item common to the school children cases (Fukushima *et al.* 1999).

In 1996 an outbreak in the USA case control studies implicated consumption of mesclun lettuce. The lettuce came from one power-processor. Cattle near the lettuce fields were excreting O157. *E.coli* was cultured from wastewater and finished lettuce. Forward training of places receiving the lettuce identified further cases.

Food and drinking water outbreaks – food outlets

Some outbreaks have been associated with particular food premises, often without a single food item being implicated. In such outbreaks there are clearly problems in determining the source of the infection. This is illustrated by an outbreak involving sixteen symptomatic cases, in September 1990, in a restaurant in Lothian. The restaurant served about 400 meals per day. Five of these cases, however, had only a tenuous link with the restaurant by virtue of their household contacts having dined there. Furthermore, the household contacts in question were all adults. In support of the investigators' conclusions, one of these asymptomatic carriers was subsequently found to be excreting VTEC O157. Nevertheless, it is not possible to say whether he passed it to the index case of the house (who was 8 years old) or whether she transmitted it to him. The investigators concluded that it must be more than a coincidence that eleven cases ate at the same restaurant even though a single food item could not be identified

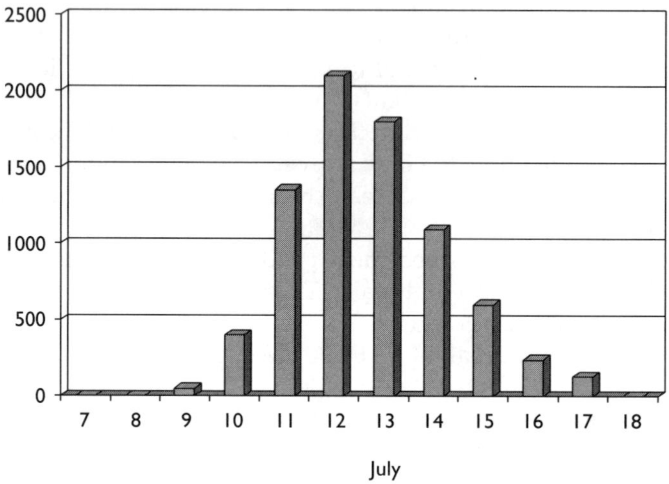

Figure 5.2 Onset of illness in children in Sakai City Japan outbreak of *E.coli* O157 July 7–17, 1996

and no isolates were made at the restaurant. The report does not indicate whether the restaurant used local suppliers of food that may also have been available at other outlets in the area and that could therefore account for other cases. Furthermore, despite active case finding only sixteen cases were identified and, of these, 25 per cent had HUS. This very high proportion of HUS cases suggests that many more cases remained unrecognised. Finally, the very long delay postulated (up to 14 days) between eating at the restaurant and illness for some of the cases has not been recorded in other point source outbreaks. Water samples and surface swabs taken at the restaurant were negative. No obvious food hygiene errors were reported although cross-contamination is the most likely explanation for no single food item being implicated (Marsh *et al.* 1992).

In Canada, in an outbreak in a nursing home, no single food item could be specifically associated with disease. One meal, which consisted of ham, turkey and cheese sandwiches, was suspected. The report does not state whether the cheese was made from unpasteurised milk. Food handling errors were recorded including the use of a scarred wooden chopping board for sandwich preparation, and storing the prepared sandwiches at room temperature. Considerable disruption to the kitchen, which was being moved, and the associated water disconnection would almost certainly have increased the opportunities for cross-contamination. The food handler had recently recovered from diarrhoea, but the report does not state whether a faecal sample from him/her was examined. All water, food and environmental samples were negative (Carter *et al.* 1987).

By 1994, more sophisticated typing methods (PFGE) were used to identify a cross-border community outbreak in central Scotland. Again, foodborne transmission was considered the most likely cause, but no particular food item could be identified. The initial cases had all eaten products from a chain of butcher's shops (one worked there) and both raw and cooked meats were implicated. Furthermore, there was evidence that poor hygiene procedures at the butcher's shop could have led to cross-contamination of cooked products. However, a later case control study of only half of the eligible cases (and this did not include the initial cases) showed an association between infection and consumption of frozen burgers from a single supermarket chain. The environmental and epidemiological evidence indicated, therefore, two routes of transmission whilst the microbiological evidence suggested a single strain was responsible for both outbreaks. No link between the local chain of butcher's shops and the frozen beefburgers could be found except that they shared single premises in a 'shop-within-shops' arrangement. Cases and controls may therefore have found it difficult to remember which they had visited and resulted in a failure to find a clear epidemiological association with a single premises, let alone a single food item. VTEC O157 was not isolated from surface swabs nor samples of meat at the butcher's shop. A sample of meat purchased from the implicated butcher's shop by one of the cases, and stored in a domestic freezer, proved negative. Burgers from the supermarket chain were also negative (Davis and Brogan 1994).

The largest outbreak of VTEC O157 in the United Kingdom to date occurred in Scotland, in November 1996. By 13 December 1996, 404 cases (of which 223 were bacteriologically confirmed) had been recognised (Anonymous 1996). By July 1997, twenty people had died. All of the clinical isolates were indistinguishable on PFGE fingerprinting. The epidemic strain was 'phage type 2 and produced VT2 but not VT1. Most cases were reported in Lanarkshire although cases were also reported from Forth Valley, Greater Glasgow and Lothian. The epidemiological outbreak investigation comprised a cohort study of approximately 100 (mainly elderly) people who attended a church luncheon on 17 November 1996. Cases were also associated with a birthday party held on 23 November. The epidemic strain was isolated from gravy which was served at the church luncheon. *E.coli* O157 was also isolated from a 6 lb piece of unopened vacuum-packed cooked beef, prepared by the suspect butcher, but the results of PFGE analysis have not been reported for this isolate. Secondary, person-to-person transmission was also suspected. The implicated butcher had a substantial wholesale and retail trade involving the production and distribution of raw and cooked meats and bakery products. The onsets of outbreak cases spanned a month and it was suspected that the latter ones were a result of person-to-person transmission ((The) Pennington Group 1997).

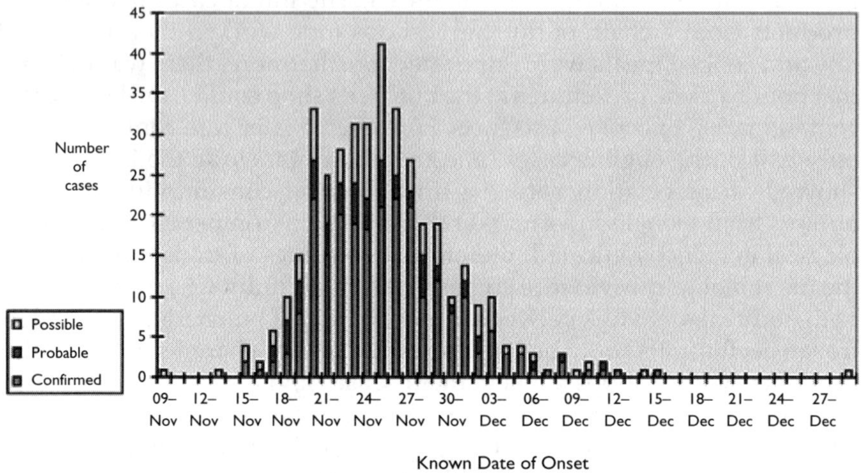

Figure 5.3 *E.coli* O157 Central Scotland Outbreak Epidemic Curve by Date of Onset of Diarrhoea

The following points are particularly relevant to environmental health:

1 The task of outbreak management and control was hampered by the complex chain of meat and meat products supplied by the butcher concerned, which had to be unravelled from the records. Clearly, accurate records of outlets supplied would have speeded up this process and aided control of the outbreak.
2 Despite significant capital investment, the layout and design of the implicated butcher constrained measures to allow effective separation of cooked and raw products. There is a need for proprietors of food businesses to seek environmental health advice prior to refurbishment.
3 The premises concerned were exempt from Meat Products (Hygiene) Regulations 1994 on the grounds that they were not handling meat products for sale other than to the final consumer.
4 It is believed that person-to-person transmission lengthened the outbreak presenting an ongoing danger in the community. The role of EHOs in preventing person-to-person transmission has already been discussed.
5 A high case fatality rate was recorded. The deaths occurred amongst the elderly who attended the church luncheon, highlighting the vulnerability of this sub-group of the population.

Some outbreak reports from England and Wales have demonstrated an association with particular food outlets, but not with a particular food item: however, opportunities for cross-contamination have been identified. In these reports the exact opportunities for cross-contamination were often not described and the possible source of infection (e.g. raw meats) not indicated. However, in some outbreaks associated epidemiologically with eating pre-cooked meats from a single outlet, opportunities for cross-contamination were obvious. For example, meats were stored in close proximity to raw beef, raw meats were left lying on preparation surfaces, and the same staff serving cooked and uncooked meats.

A lack of success in isolating VTEC O157 from food and environmental samples during outbreaks was reported by Wall *et al.* (1996) in their review of eighteen general outbreaks of VTEC O157 between 1992 and 1994. In the four English outbreaks, where foods or environmental samples were submitted, VTEC O157 was isolated once. The PFGE method, for accurately differentiating strains of VTEC O157, has revolutionised the investigation of such outbreaks. Now, extraneous cases can be omitted from the analysis and strong associations with particular supermarkets have been demonstrated.

In August 1993, a cluster of cases was recognised in South East Wales. An initial association with various meats from a single butcher's shop was discovered. Enquiries at the butcher's shop revealed opportunities for cross-contamination, but all food and environmental samples taken at the butcher's shop were negative for VTEC O157. Some weeks later, a frozen beefburger, purchased at the time of the outbreak, and stored in a customer's domestic freezer, yielded a strain of VTEC O157 which was identical to the outbreak strain. This represented the first isolation of VTEC O157 from a food in the UK (Willshaw *et al.* 1994).

A butcher's shop was the setting for a further outbreak of HUS and haemorrhagic colitis in November 1995. The epidemiological evidence for an association with the butcher's shop was not that convincing as a definite link with the shop could only be established for six of the eleven cases and three had eaten cooked ham. However, environmental health findings at the shop certainly supported the hypothesis that it was the source of the outbreak although samples of food and environmental swabs were negative. The supporting environmental evidence was summarised as follows: 1 Extensive production of cooked meats was being carried out in limited space alongside the preparation of raw meats; 2 the same staff handled raw and cooked meats; 3 in the period leading up to the outbreak, hams were being sealed by hand (as opposed to being vacuum packed) which could have allowed water from the cooling tubs to enter; 4 breadcrumbs were applied manually to the cooked hams providing further opportunities for contamination; 5 as the vacuum sealer was broken, hams were packed by hand, providing an aerobic atmosphere favouring the growth of VTEC (Gammie *et al.* 1996).

Food and drinking water outbreaks – drinking water

Further community outbreaks have been reported where no microbiological evidence of the source could be found. In 1990, thorough microbiological analysis of foods and water was carried out as part of an investigation into a cluster of four cases in a village in Grampian. No isolates were made. Nevertheless, the investigators concluded that the village potable water was contaminated with cattle slurry after a subsidiary supply had been used to meet demand in hot, dry weather. Intermittent contamination was given as the explanation for so few cases in a village of 1,700 people (Dev *et al.* 1991, 1991a). More recently, a change in weather conditions (prolonged dry spell then heavy rain) coincided with bacteriologic failures of private water supplies in Grampian. The rise in VTEC O157 infections was thought to be due to slurry on the fields being washed into drinking water supplies (Curnow 1994). In 1995, VTEC O157, PT2, was isolated from contaminated drinking water in Scotland (Thompson-Carter *et al.* 1996).

In the USA a large outbreak (243 cases, 21 confirmed) affecting residents of and visitors to a single town was attributed to consumption of the unchlorinated municipal water supply. Seepage from sewer lines and overflows located near the water lines were the suspected source of contamination. The sewage was able to enter small leaks in the water lines when repairs at the water meters caused a back flow and loss of positive pressure (Swerdlow *et al.* 1992).

A more recent outbreak of VTEC O157 and *Campylobacter* was attributed to a public water supply in Fife, Scotland. The water supply became contaminated by stream water. The sewage outflow for the village was upstream of the point where the stream water had been abstracted. In all 765 residents and workers reported illness; VTEC O157 was isolated from six of them. Boiling notices were issued and bowsers provided at locations in the village (Jones and Rowarth 1996).

To conclude, several sources of foodborne outbreaks have been reported. Foodborne outbreaks have been reported in the community, in day-care centres and in care institutions. The source of the initial contamination is usually food of bovine origin (i.e. raw beef or milk). These sources act as the vehicles of transmission if they are not adequately heat treated (e.g. undercooked beef or raw milk and cheese, salami which is not heat treated). Sometimes the beef is accidentally undercooked (e.g. with a grill not working properly) or it may be desired 'rare'. Such products also act as a source of contamination to cooked products. Other food products can be contaminated by bovine faeces (e.g. potatoes and apple juice).

Once milk is pasteurised it can be cross-contaminated by raw milk or by bovine faeces. The acidity of some foods (e.g. produced by a fermentation process in the case of salami and apple juice) cannot be relied upon to kill

VTEC O157. Cross-contamination of cooked products is a further feature. Outbreaks have been identified where several products from one premises are associated with infection indicating extensive cross-contamination. Therefore the critical control points for VTEC O157 are adequate heat treatment of the raw product and prevention of cross-contamination by complete separation of raw and cooked products.

Animal outbreaks

Direct zoonotic spread from farm animals, at a farm visitor centre, was thought to be responsible for four cases of VTEC O157 in Nottingham in 1994. Microbiological investigations showed that cattle and goats at the farm were excreting VTEC O157, which was indistinguishable (by RFLP) from the human isolates. In all, six cases had evidence of VTEC O157 infection (either culture or serology). Of these, all six had handled animals, and five had washed their hands, despite the reported inadequacy of hand-washing facilities. Of course, the thoroughness of hand washing could not be guaranteed. Also, the cases may have eaten or put their hands in their mouths before washing their hands and, as most of them were under 7 years of age, this is likely. Opportunities existed for close contact with animal faeces, as visitors were encouraged to enter the animal pens. The outdoor picnic area showed signs of faecal contamination by livestock. Nevertheless, the report does not state that any of the cases recalled obvious gross contact with animal faeces. Hence, contact would have been with only small quantities of animal faeces. The case finding involved clinical surveillance of HUS and laboratory screening of all specimens. The high proportion of cases with HUS (50 per cent) suggests that less serious cases were missed (Shukla *et al.* 1995).

Detailed case control studies can further identify on-farm risk factors for outbreaks. In one outbreak where fifty-one cases had visited a dairy farm a case control study showed that cases were more likely to have contact with cattle, to bite their nails and to purchase food from an outdoor stall. Those who washed their hands before eating were at lower risk. It was estimated that 7,000 of the 75,600 people visiting the farm during the outbreak may have developed diarrhoea as a result of their visit.

Particularly high levels of hygiene, i.e. thorough hand washing, should be maintained where members of the public are encouraged to touch animals. However, even where individuals have not actually touched animals they should be encouraged to wash their hands before eating or putting their hands in their mouths. The ease with which VTEC O157 can be contracted from the general farm environment lends further weight to the argument that the organism has a low infectious dose.

Person-to-person spread outbreaks

VTEC O157 outbreaks attributed to person-to-person transmission by the faecal–oral route have been reported in institutions where individuals cannot exercise a high level of personal hygiene. Outbreaks have been reported in psycho-geriatric wards, homes for the elderly and children's day-care facilities. In children's day-care centre outbreaks, 3 per cent to 38 per cent of children have been recognised as cases (with case definitions including bacterial confirmation, bloody diarrhoea or HUS). The introduction of VTEC O157 into a children's day-care centre (pre-school age children) can transform a small community outbreak into a major incident. Infection is usually introduced by a sick child. Detailed record keeping of toileting, particularly in young children, facilitates early detection and control of such outbreaks. Attack rates were higher in younger ages. In some outbreaks, in such premises, the initial vehicle of infection was food with subsequent extensive person-to-person transmission.

Secondary transmission has also been reported in community outbreaks. In the large beefburger-associated outbreak in the USA in 1993, it was estimated that 11 per cent of the identified cases were infected by secondary, presumably person-to-person spread. Person-to-person transmission has also been reported in hospitals.

Person-to-person transmission of VTEC O157 is clearly a risk to public health. The disease has a low infectious dose which allows it to spread easily from person-to-person especially where personal hygiene is poor. Nevertheless, it can also be transmitted in private homes and this leads to a greater number of people becoming ill in outbreaks. Person-to-person transmission is a particular danger in children's day-care facilities where the under 5-year-olds are particularly at risk of developing HUS. In such outbreaks, younger children appear to be most at risk of contracting the infection. An important control measure is preventing children with diarrhoea attending such premises. Adequate hygienic precautions are also important.

Outbreaks from recreational exposures

Community outbreaks have been associated with person-to-person transmission of infection, e.g. via swimming pools where inadequate disinfection procedures are carried out. Faecal contamination of the swimming pool by an infected individual was thought to be the most likely source.

Swimming in freshwater lakes, faecally contaminated by bathers, was the source of a further outbreak, where *Shigella sonnei* was also isolated from cases. The involvement of *Shigella sonnei*, which unlike VTEC O157 does not have an animal host, makes it more likely that the source of the contamination was human faeces rather than animal manure. Swallowing the water, and spending longer in the water were associated with increased risks of infection.

To conclude, an infected individual can present a risk to others by contaminating both natural bathing waters and swimming pools. In swimming pools, adequate chlorination must be maintained. Members of the public must be discouraged from defecating in bathing waters. Animal manure from surrounding agricultural land is a further potential source of contamination in natural recreational waters, e.g. rivers and lakes.

A practical approach to the investigation of an outbreak

Identification and confirmation of the problem

Checking information

All epidemiological data should be confirmed by independent enquiry as thoroughly and quickly as possible before action is taken. These data should be accurately recorded and retained for future reference (including possible use as legal evidence).

Checking diagnosis

The possibility of laboratory mistakes, and of equivocal results with some laboratory techniques, requires that laboratory methods should be reviewed and, where appropriate, isolations confirmed by a reference laboratory, which may also be able to help with typing. Isolates and sera should be saved for further typing.

Confirming an outbreak exists

Changes in diagnostic techniques and clinical habits and population structure can cause artefacts in surveillance data, and the media can cause inappropriate public alarm. An increase in investigations due, for example, to a new interest or concern of clinicians should only increase the total number of samples submitted. When a problem has been confirmed, interim control measures to be instituted should be considered. At this stage, environmental samples (water, food) that may be required later should be taken.

Communication and designation of responsibilities

It is essential that everyone who needs to know about an acute incident is given early accurate information and knows what tasks they are expected to perform.

Patient interviews

In large outbreaks, it will not be practical to interview all cases in depth. A small number of cases (5–10) should first be interviewed in depth to explore possible exposures. It may be helpful to interview a few unaffected persons from the at-risk population as well as any atypical cases or persons with very limited exposures. From these interviews, the population at risk will be defined, and hypotheses of the cause of the outbreak will be generated. An appropriate questionnaire can then be designed.

Defining population at risk

Often this will be obvious: those attending a single function or reception or those living in an institution. Less easy to define are those outbreaks occurring in the general community. Definition of the population is essential in epidemiology, since all investigations and results should be referable to that group and as a preliminary step for case-searching.

Case-searching

Cases presenting at the point of recognition of an outbreak are seldom typical. Vigorous case-finding will be necessary to measure the size of the outbreak, and to increase the power of any proposed case control study. A preliminary step is to decide on a case definition. This may be approached in two stages. In the first place, a broad case definition is used; laboratory investigation is then used to reclassify possible cases as confirmed or unconfirmed. Methods of case-finding, both retrospective and prospective, include:

- review of medical records (GP, hospital, occupational)
- soliciting clinical reports (GPs, hospital doctors, nurses) and statutory notifications
- review of laboratory data
- survey of population.

Analysis of case data and formulating a hypothesis

Data should be analysed by time, place, person and by cross-tabulation, keeping an open mind since new vehicles and new modes of transmission may be identified.

Hypothesis testing

Hypothesis may be tested:

- epidemiologically by case control or cohort studies

- microbiologically (testing food samples)
- by searching for environmental and other data, e.g. history of water pollution, engineering data or catering practices.

Analysis and presentation of data

Data from patients must be summarised and compared in order to identify possible common exposures or risk factors.

Time

The time of crucial importance is the time of onset of disease, since from this and a knowledge of the incubation period of the infection the period of exposure to infection can be determined. In an outbreak, the dates of onset in cases should be presented graphically, usually in the form of a histogram.

The epidemic curve is the most useful and immediate means of assessing the type of outbreak. In point-source outbreaks in which all cases are exposed at a given time (such as a wedding reception), onset of symptoms of all primary cases will cluster within 3 to 8 days of a common exposure. An epidemic curve that extends beyond a single incubation period suggests either a continuing or recurring source of infection, or the possibility of secondary transmission.

Place

Cases clustering in a neighbourhood may indicate the existence of a point-source of infection or of person-to-person spread.

Person

Analysis of patient characteristics and histories of exposures (age, sex, medical history, occupation, travel history), may suggest hypotheses about the source and mode of infection.

Case control and cohort studies

Descriptive data may suggest hypotheses about the source of mode of transmission of an infection, but they are not always a sufficient basis for introducing control measures. Observations on the cases have to be set against what is expected in the population from which they come. Analytical methods are required to test the hypotheses generated from descriptive data. Epidemiological evidence is necessary to demonstrate an association between the contaminated food and the occurrence of illness. Essentially the epidemiological approach is to compare the characteristics of infected persons (cases) with those of a similar group of uninfected persons (controls). Two principal study designs are used, the case control and the cohort. Choice of method in practice is usually determined by what is feasible. Where a cohort can be readily identified, it is usual to undertake a cohort study.

The essential difference between a case control study and a cohort study is that the case control study begins by identifying people with and without the infection and then retrospectively tries to identify factors associated with disease. In cohort studies, groups of people are identified other than by the presence of disease, and then information on disease occurrence is sought. Cohort studies may be prospective, when the disease occurs after the study has begun and the population characteristics have been identified; or retrospective, as is usually the case in investigation of acute incidents. The attack rate in persons eating the food (exposed) is compared with those not eating the food (not exposed). The method can be applied to any hypothesised vehicle of infection and to hypothesised risk factors (factors which are not necessarily the source of infection but contribute to the likelihood of infection). In addition to the factors also relevant to case control studies the following need to be carefully considered.

The descriptive epidemiological enquiry will reveal the population at risk for the infection. Thus, in the previous example, interviews with the first cases of food poisoning revealed that the population at risk comprised delegates attending a particular conference. In outbreaks where a specific at-risk group cannot be identified, the case control approach may be more appropriate. If the at-risk population is large, a sample can be investigated and results extrapolated to the total population. In order to do this the sample must be representative. Ideally a random sample should be taken by applying random number tables to the sampling frame. In outbreaks in institutions, a complete list of the population can readily be obtained. An alternative to random number sampling is to select every nth person from the list. Cluster or stratified sampling may sometimes be appropriate. In both types of study the basic analysis is by a comparison of proportions.

Why are VTEC O157 outbreaks different from other food poisoning outbreaks?

1 The potential for hospitalisation, serious disease, permanent disability and death is higher, therefore the outbreak will attract more media and public interest. There is a need for effective risk communication to stop panic and to present accurately public health information.
2 Secondary spread from outbreak cases to their contacts is a real possibility and satellite outbreaks, in e.g. children's day care, could occur.
3 Tracing cases is more important because they may be in an occupation where they could further spread the disease, e.g. foodhandlers.
4 Cases of HUS may not be identified until several weeks after infection.

Summary

Outbreaks of VTEC O157 have been attributed to:

- Contaminated beef (particularly ground/minced) beef which was consumed raw (salami) or undercooked (rare beefburgers).
- Unpasteurised milk or dairy products and pasteurised milk/dairy products contaminated post-pasteurisation.
- Fruit and vegetables (or products made from them) which have not been cooked/pasteurised.
- Cooked sliced meats which have been contaminated after cooking. [Outbreaks associated with particular food businesses but not with any one food item have also been attributed to cross-contamination.]
- Unchlorinated drinking water supplies or chlorinated supplies contaminated after chlorination.
- Direct zoonotic spread via contact with animals or their manure.
- Person-to-person transmission via faecal–oral spread particularly in situations where strict personal hygiene is difficult to maintain, e.g. children's day-care centres.
- Recreational water contaminated by human or animal faeces.

KEY MESSAGES

- VTEC O157 outbreaks can be caused by contamination of food, water and the environment via numerous transmission routes. Treating food and water with heat/chemicals does not preclude post-treatment contamination.
- Outbreaks should be investigated promptly and systematically.

Chapter 6

Animal and environmental reservoirs of VTEC

Evidence from outbreaks amongst humans provides strong evidence that the major reservoir of infection for humans is cattle. This has led to a number of surveys of this organism in cattle and other animals as well as in abattoirs. VTEC rarely causes disease in animals because it is usually carried as inapparent infection and therefore veterinary reports of disease in animals are not a sufficient basis for assessing public health risk.

Farm animals

Evidence available to the ACMSF working group on VTEC (ACMSF 1995), which reported in 1995, showed considerable evidence that farm animals were an important reservoir of infection in the UK. In addition, they recorded that the government should fund research programmes to establish the incidence and prevalence of *E.coli* O157:H7 in cattle and other livestock and identify the husbandry and other factors contributing to herd infection and control.

A study of *E.coli* isolates from diseased farm animals in England and Wales from 1986 to 1991 found that 2.8 per cent of bovine isolates, 6.1 per cent of ovine isolates and 4 per cent of porcine isolates produced vero cytotoxins (Wray 1993). However, the porcine isolates were mainly serogroups O138, O139 and O141 which are not associated with human disease. Investigations in Yorkshire by Chapman and others found that O157 was isolated from 84 (4 per cent) of 2,103 bovine rectal swabs taken immediately prior to slaughter, and from 30 (7 per cent) surface swabs of carcasses taken from cattle positive on rectal swabs. Meat from this slaughter house had been linked with several human cases of VTEC O157 in the North of England, and so these results could not be regarded as representative of the national herd (Chapman *et al.* 1993). Nevertheless, VTEC O157 was also isolated from 2 (8 per cent) of 25 carcasses of rectal swab negative cattle which indicates that carcass-to-carcass contamination during the slaughtering process is a route of infection for carcasses from uninfected cattle.

In order to establish the possible sources of human VTEC infections in Cornwall cases were interviewed about farm exposures (Trevena *et al.* 1999). From November 1994 to October 1997, four laboratories identified sixty-nine cases in humans, and twenty-two farm investigations were carried out in which animal rectal swabs, faeces specimens, pre-stream milk samples and various environmental samples were tested. In nine investigations, VTEC O157 was found and in seven the strain was indistinguishable from the human isolation by phenotypic and genotypic subtyping. Cases included farm visitors, holidaymakers, farming families and farm workers.

Studies in abattoirs confirm the high rate of contamination of cattle. The UK wide study of beef carcasses in 1994/5 recovered VTEC from 0.47 per cent of carcasses. In a 1-year study in Sheffield in 1995/6 VTEC was isolated from 15.6 per cent of cattle in abattoir and 2.2 per cent of sheep, but not in 1,000 pig rectal faecal samples. Further studies have now been conducted at a UK level. In the cattle and sheep survey, faecal samples taken from animals sent to slaughter throughout 1999 were positive for VTEC O157 in 4.7 per cent of cattle, and 1.8 per cent of sheep. Faecal carriage was detected throughout the year but was greatest in the summer in sheep and cattle and lowest in the autumn in cattle and in the winter in sheep. In summer, faecal contamination rates in cattle reached 5.5 per cent, and in sheep 2.4 per cent. These rates fell to 4 per cent in cattle in autumn, and to two (0.9 per cent) in sheep in winter (Paiba 2000). Prevalence was highest in cattle in abattoirs in the east of the country. In the pig survey only 0.16 per cent of animals were positive.

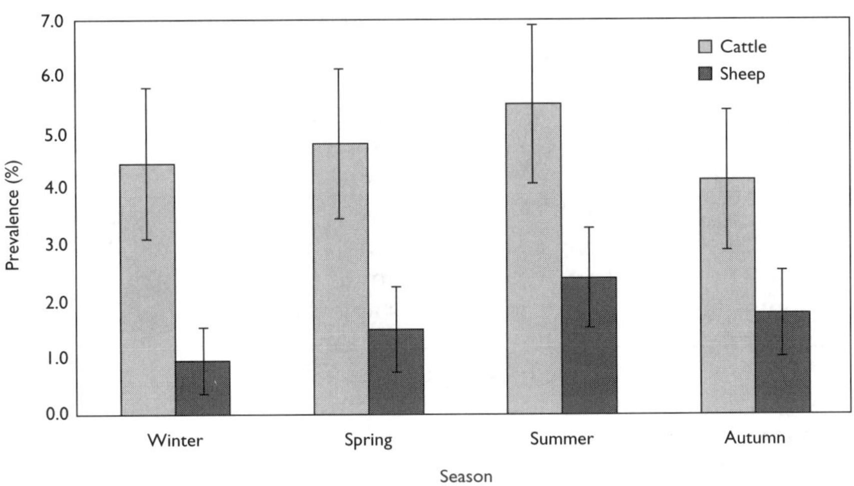

Figure 6.1 Histogram of the prevalence of faecal carriage of VTEC O157 by season for cattle and sheep

Further on-farm studies have been conducted to determine the presence of faecal excretion by cattle on randomly selected farms in England and Wales (Paiba 2000). VTEC was isolated from 4.7 per cent of individual animals and 44 per cent of herds between June and December 1999. In affected herds 10.2 per cent of individual cattle were excreting VTEC. Excretion was significant with age and greatest in cattle under 24 months. Cattle in fattening herds are twice as likely to be excreting than those in dairy or suckler herds. These studies, though carefully done, did not use the most sensitive techniques for identifying VTEC and therefore further studies with more sensitive techniques may well demonstrate higher carriage rates amongst animals.

Major studies are going on in Scotland to identify the source of the high incidence of human cases there (Synge *et al.* 2000). In one study beef cattle aged 12 to 30 months on 952 farms were sampled randomly around Scotland; 14,564 samples were collected between 1998 and 2000; 8.6 per cent of animals in the sampling frame were found to be shedding *E.coli* O157, and 23.7 per cent of herds sampled had at least one shedding animal. Dairy, beef and mixed farms had similar rates of positivity. Isolation rates were highest amongst animals that were housed rather than grazing on grass. Those animals that were housed had higher rates of shedding *E.coli* in the spring and lower rates in the winter. The proportion of positive farms varied by area in Scotland. It ranged from 26 per cent positive in the North East to 17 per cent in the Highlands. A follow-on study of shedding beef suckler cows showed that most positive farms shed for less than 5 months of the year. Shedding was higher in spring and autumn and possibly associated with type of feed.

The recent UK studies give faecal carriage and excretion rate in cattle, sheep and pigs of similar levels to results reported in other parts of the world.

Considerable work has been undertaken in the USA and Canada. In a 14-month longitudinal study on four dairy farms in Wisconsin, fifteen heifer calves from each farm were sampled weekly from birth. Two farms remained negative from VTEC but one farm had two separate periods of *E.coli* O157:H7 shedding lasting 4 months, and 1 month and the fourth farm had at one positive and for a 5 month period. Individual heifers shed strains in faeces from 1 to 16 weeks at levels ranging from 2.0×10^2 to 8.7×10^4 CFU per g. Environmental sampling revealed the same subtypes in other cattle, feed, flies, pigeons and water (Shere *et al.* 1998). This study demonstrated that strains introduced into farms can persist in a herd for 2 years.

A 1999 survey in the USA of the prevalence of VTEC O157 in faeces and on hides presenting for slaughter at abattoirs found higher rates than previously estimated. Early estimates suggested that fewer than 10 per cent of cattle carried the organism, but the study found that 28 per cent of cattle were faeces positive, 11 per cent of hides were contaminated and that 43 per cent of carcasses sampled before evisceration were contaminated. After processing only 1.8 per cent were positive (Elder *et al.* 2000).

The ability of VTEC O157 to survive in bovine faeces is very relevant to its ability to contaminate foods grown on contaminated pasture. In one experiment, *E.coli* O157 could be isolated from a manure pile collected from experimentally inoculated sheep for 21 months. Bacteria survived in manure frozen at –20°C for at least 100 days. Studies in Scotland have been carried out of the survival of *E.coli* in slurry applied to different types of land. *E.coli* could be recovered from soil at 100 per cent success rate for the first 3 days after application. At 29 days only 1 per cent of samples were positive. *E.coli* O157 did not behave differently from other *E.coli* strains.

Food and water

Several techniques to isolate VTEC from food have developed in recent years. The advent of immunomagnetic separation (IMS) in 1994 significantly increased the isolation rate of O157 from food. The PHLS has developed a standard method for food products. The principle involves selective enrichment with isolation from the enriched culture by IMS. The selective agar cultures are then examined for characteristic colonies. *E.coli* O157 is confirmed by serological and biochemical tests.

VTEC organisms grow well under aerobic or facultatively anaerobic conditions. They live happily in the intestinal tract of humans or animals and can multiply to large numbers. It has been shown that O157 can survive for long periods in water and it is likely that environments contaminated with faecal material can be sources of infection. In food, O157 strains are killed by pasteurisation and normal cooking temperatures. The organisms survive well frozen and can be recovered after many months.

Several studies have examined the survival characteristics of VTEC O157 in foods, particularly in relation to techniques used for food preservation. Resistance to acidic conditions can be important to foodborne pathogens for survival in the gastrointestinal tract and in acidic foods. Once ingested the pathogen first encounters the highly acidic environment of the stomach (gastric acid). This is followed by the less acidic environment of the small intestine. Available evidence suggests that organisms with mechanisms for inducing acid resistance are better equipped to withstand these acid challenges. Furthermore, it is suggested that the infectious dose of enteric pathogens is related to their ability to withstand the environment of the gastrointestinal tract with the more acid-resistant organisms having the lower infectious dose.

Acidic foods such as apple juice have in the past been regarded as low risk in terms of enteric pathogen transmission, but surprisingly several outbreaks of VTEC O157 associated with such foods have been reported.

Three mechanisms of acid resistance have been described for *E.coli*, namely, oxidative, glutamate-dependent and arginine-dependent systems. Experiments have shown that, in *E.coli* O157, these acid resistance systems could significantly contribute to its survival in acidic foods and the intestine. Furthermore, it has been hypothesised that the gastrointestinal tract of cattle could induce one or more of the acid resistance mechanisms. When the bacteria are released from the gastrointestinal tract of cattle and enter either an acidic food or directly into the human gastrointestinal tract they are already acid-adapted and able to survive. It has been demonstrated that, once activated, the acid resistant system will remain active over long periods of cold storage. Hence, acid resistant VTEC O157 contaminating beef carcasses will remain resistant in refrigerated carcasses.

Experiments have demonstrated that VTEC O157 can survive in apple juice (pH2) for up to 21 days, and that lower storage temperature (4°C) enhanced survival at adverse pH. Some strains of *E.coli* O157 survived for 24 hours at pH2 (Garland Miller and Kaspar 1994). VTEC O157 also survived for more than 50 days in mayonnaise-based dressing stored at 5 to 7°C (Erickson *et al.* 1995), and in ground beef and mayonnaise salads stored at 5°C for 72 hours. Indeed, there were significant increases in populations of VTEC O157 in such salads stored at 21°C and 30°C (Abdul-Raouf *et al.* 1993). Further studies have shown that where *E.coli* O157 are initially present at levels of 10^4/g in meat, they would not all be killed during the production of fermented sausage that does not receive a pasteurisation treatment (Glass *et al.* 1992). In these experiments, the organism did not grow during fermentation but did survive for 2 months in the final product even though numbers had decreased. VTEC O157 can multiply on lettuce and cucumber stored at 21°C for up to 14 days but numbers decline on vegetables (particularly carrots) (Abdul-Raof *et al.* 1993).

The ability of *E.coli* O157 to survive heat treatment seems to depend on the medium in which it is contained. There is some evidence that it survives better in fatty minced beef than in lean minced beef. It has been suggested that in fatty foods there is less water and hence the heat is not transmitted so well as in products containing a lot of water. Cooking processes that produce an internal product temperature of 60°C for 2–3 minutes provide a >5D kill (i.e. a reduction of 10^5 organisms) of *E.coli* in a variety of meat products. Heat shock of *E.coli* O157, prior to heat treatment, resulted in an increased ability to recover after heat treatment. Also, anaerobic growth of the organism resulted in an increase in the numbers of the organism able to recover after the heat treatment (Murano and Pierson 1992). This has implications for pre-cooking of meat products that are then frozen and reheated.

Further research has demonstrated that strains of VTEC O157 survived in ground beef frozen to –20°C for 9 months with little change in numbers

(Doyle and Schoeni 1984). However, it is killed by standard pasteurisation temperatures (D'Aoust *et al*. 1988).

Several surveys have been conducted on the prevalence of VTEC O157 in food products. For example, in Scotland from 1997 to 1999, 2,429 samples of foodstuffs were examined by IMS. Specimens were collected by EHOs and included raw retail meats (n = 1,190), raw milk cheese (n = 739) and raw milks (n = 500). Only 0.24 per cent of retail raw beef products were positive (one beefburger and one beef sausage). There were no other positive foods (Coia 2000).

A LACOTS/PHLS survey conducted in May and June 1999 (Little and Mitchell 2000) examined red burgers purchased anonymously from burger outlets including takeaway and burger restaurants, mobile vendors, temporary stalls and other points of sale. Samples were transported by local environmental health staff to the laboratory in accordance with Food Safety Act and 1990 Code of Practice No 7. Information on the burger samples and outlets was obtained by observation and recorded on a standard proforma. This included how the burger was cooked and what method was used by the vendor to determine whether cooking was adequate. Sampling was anonymous so information on the outlets and practices with regard to food safety legislation was obtained from details obtained from the last inspection visit. Four 25 gram samples were taken of burger meat for examination using standard PHLS methods for food products. A total of 3,128 burger samples were obtained and examined by 46 laboratories in the UK. Fifty-three local authority food liaison groups participated in the study and samples were submitted by 366 local authorities. Of the 3,128 burger samples examined, 255 (8 per cent) of samples had aerobic colony counts of 10^4 CFU per gram or more and 29 (1 per cent) had enterobactera 10^4 CFU per gram. *E.coli* was identified at 10^2 CFU per gram or more in 3, that is less than 1 per cent of samples. Salmonella species, Campylobacter species and *E.coli* O157 were not detected in any samples examined.

A survey of VTEC in unpasteurised milk cheeses was recommended by ACMSF in 1995 because of the relative acid tolerance of the organism. British standard methods were used to examine for the presence of *E.coli* and O157 H7 using an immuno-magnetic separation method. The 801 samples were purchased from retail outlets in England and Wales during January and February 1997. The 558 samplers who took part in the survey were trained by MAFF in the collection of surveillance samples. Samples were taken to eight accredited laboratories where the 801 cheeses of 159 different varieties were examined. The cheeses were mainly from England, France, Switzerland and Italy and were purchased from grocers and delicatessens. Most samples had very low counts for *E.coli* and none were positive for O157 H7.

Summary

- The major animal reservoir of VTEC is cattle, VTEC O157 was present in the faeces of 47 per cent of cattle slaughtered in Great Britain for human consumption in 1999/2000.
- In a random sample of farms studied in England and Wales in 1999, 44 per cent of cattle herds had evidence of infection.
- Shedding is common in calves, fattening herds, and in spring and autumn.
- Sheep are also a reservoir of infection with 1.8 per cent faecal carriage at slaughter in 1999.
- Pigs rarely excrete VTEC O157 with only 0.16 per cent faecal carriage at slaughter in 1999.
- *E.coli* O157 can survive for long periods in the farm environment.
- *E.coli* O157 can withstand acid environments, such as apple juice, for many days and may be relatively heat resistant.
- Surveys of retail foods suggest that VTEC O157 is rare, even in meat products.

KEY MESSAGES

- VTEC is likely to be present in up to half of the cattle herds in the country and in the farm environment.
- Sheep are also potential sources of infection.
- Outbreaks have been caused by unexpected food vehicles because of the acid tolerance of the organism. Therefore, don't rule out foods without evidence.

Chapter 7

Control of VTEC O157

Introduction

This chapter addresses the known transmission pathways for VTEC O157 from a bovine reservoir to a human host. Critical control points have been identified within the transmission pathways. For each critical control point the legislative and practical control measures have been discussed. Also, the agencies involved in maintaining the control points are indicated. Policies for preventing human infection with VTEC O157 are formulated at international, national and local levels. For each control point, recommendations for control suggested by the following groups are reported:

- Advisory Committee on the Microbiological Safety of Food
- Task Force on *E.coli*
- The Pennington Group
- World Health Organization

Background on each of the organisations

ACMSF

The Advisory Committee on the Microbiological Safety of Food (ACMSF) was set up by UK Health and Agriculture Ministers in 1990 in response to a recommendation in the report of the Richmond Committee on the Microbiological safety of Food. Its terms of reference are:

> To assess the risk to humans of micro-organisms which are used, or occur, in or on food, and to advise Ministers on the exercise of powers in the Food Safety Act 1990 relating to the microbiological safety of food.

Membership comprises academics, independent consultants, industry, consumers and enforcement authority. The secretariat is the Food Standards agency.

In 1992, ACMSF set up a Working Group on vero cytotoxin-producing *Escherichia coli*. The terms of reference of the Working Group were:

> To assess the significance of VTEC as a foodborne pathogen and to advise on any action which could be taken to reduce foodborne disease associated with it.

The Working Group published its report and recommendations in 1995. Subsequent revised recommendations have been published as part of ACMSF annual reports.

Task force on E.coli *O157*

Joint Food Standards Agency / Scottish Executive

The Task Force on *E.coli* O157 was appointed under the joint sponsorship of the Food Standards Agency Scotland and SE Health Department to:

> Review the risk to health of the public in Scotland, and current activities to prevent human infection with *E.coli* O157, assess the effectiveness of the present arrangements for co-ordination of action at national and local level and consider what future measures would help protect the public. The team includes expertise in epidemiology, public health, veterinary medicine, microbiology, agriculture, water supply, environmental health and consumer interests. A final report is expected in June 2001.

(The) Pennington Group

As part of the government response to the largest UK outbreak of VTEC O157 that occurred in Scotland in November 1996, an expert group under the Chairmanship of Professor Hugh Pennington was established. The Pennington Group, which sat from the beginning of December 1996 to the end of March 1997, was asked to 'examine the circumstances which led to the outbreak in the central belt of Scotland and to advise on the implications for food safety in general and the general lessons to be learned'. The Group also took into account further outbreaks in Tayside, the Borders and Lothian, which were identified during the enquiry. The Group published their interim report and priority recommendations on 15 January 1997 and a final report on 8 April 1997.

WHO consultation on the prevention and control of enterohaemorrhagic* Escherichia coli *(EHEC)

The objectives of the consultation were to share experiences and lessons learned regarding the contamination of the food supply with EHEC strains, and to exchange information on current approaches to the prevention and control of infections and outbreaks caused by *E.coli* O157:H7 and other EHEC, in addition to developing recommendations for surveillance, outbreak investigation, control and prevention, and research needs.

CONTROL POINT 1: ON-FARM CONTROL

Aim: To eradicate or minimise carriage or shedding of VTEC O157 by bovines

Agencies: MAFF, Veterinary Laboratory Agency, farmers' unions and representative organisations (e.g. NFU).

Published UK guidance/recommendations

ACMSF: Recommendation for research into epidemiology of VTEC O157 in agricultural livestock and husbandry and other factors contributing to herd infection and control.

Pennington: Education/awareness programme for farm workers, repeated and updated periodically to ensure that they are aware of the existence, potential prevalence and nature of *E.coli* O157.

Evidence presented to VTEC taskforce: Investigate measures to reduce contamination of animal feed, pastures and water courses by wild birds: investigate measures to reduce direct contact between wild birds and livestock.

WHO: Research recommended in the following areas: Determine prevalence of EHEC in wild animals, insects, birds, fish and shellfish, pets as well as livestock.

Determine sources and transmission of EHEC contamination of livestock so as to target intervention/prevention strategies regarding:

- survival in water, soil and feed (including silage), manure/slurry
- transfer from mother to calf, faeces to hide and human to animal
- decontamination of sewage and the use of slurry in agricultural practice.

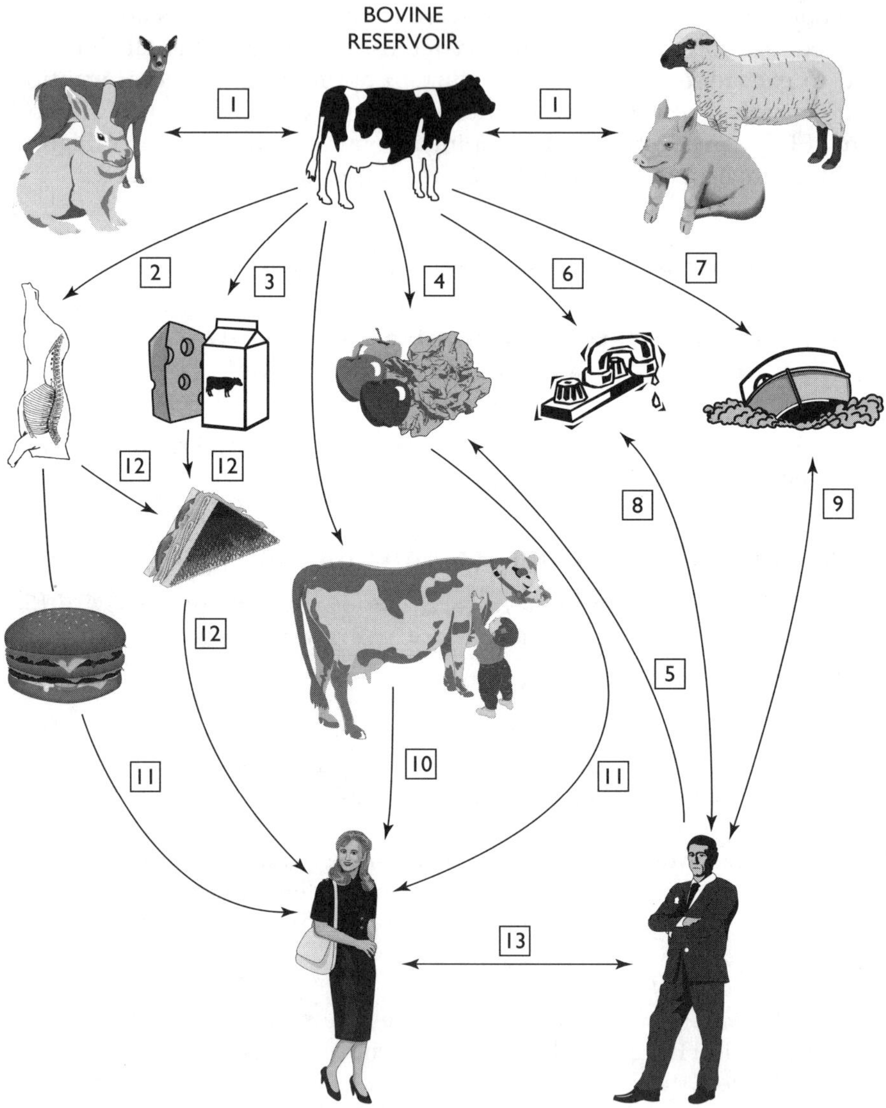

Figure 7.1 Transmission of VTEC O157

Determine factors which influence colonisation and excretion in farmed animals including transportation, diet, immune response, growth promoters and antibiotics.

Figure 7.2 Control Point 1: On-farm control

Evaluate methods to reduce the presence of EHEC in farm animals, e.g. the use of colicin-producing organisms, competitive exclusion, VTEC-specific 'phage.

Key points and practical issues

- Cattle herds (dairy and beef) are considered to be the reservoir for VTEC O157 in the UK. Whilst infection in individual animals is transient it can recirculate in a herd primarily through manure-contaminated pasture and animal drinking water supplies. It can survive in such environments for at least several months.
- Sheep, pigs, goats, deer and horses can also carry VTEC O157 and may play a role in spreading infection between herds or propagating infection within one farm. Wild mammals and probably flies and wild birds also disseminate contamination within and between herds.
- Given that VTEC O157 is now clearly prevalent in both wild and domestic animal populations total eradication of VTEC from bovine herds is unlikely. The use of antibiotics in animals used for food production is not without its risks and is controversial. Research into the use of probiotics and competitive exclusion is still in the early stages. Meanwhile, it may be possible to limit the circulation of the bacteria within a herd by changes in farm management practice. Possibilities include treating animal drinking water supplies; regular cleaning of animal drinking troughs to remove manure; allowing time to elapse between spreading slurry on grazing ground and allowing animals access. Additionally, there may be factors that increase shedding of VTEC O157 by an infected animal, e.g. age, condition, diet and stress (e.g. at transport) and which could be controlled by animal husbandry measures.

Further reading

Chapman, P.A. and Ackroyd, H.J. Farmed deer as a potential source of verocytotoxin-producing *Escherichia coli* O157 [letter]. *Veterinary Record* 1997; 141: 314–15.

Gagliardi, J.V. and Karns, J.S. Leaching of Escherichia coli O157:H7 in diverse soils under various agricultural management practices. *Applied & Environmental Microbiology* 2000; 66: 877–83.

Garber, L., Wells, S., Schroeder-Tucker, L. and Ferris, K. Factors associated with faecal shedding of verotoxin-producing Escherichia coli O157 on dairy farms. *Journal of Food Protection* 1999; 62: 307–12.

Hancock, D.D., Besser, T.E., Rice, D.H., Herriott, D.E. and Tarr, P.I. A longitudinal study of *Escherichia coli* O157 in fourteen cattle herds. *Epidemiology & Infection* 1997; 118: 193–5.

Hancock, D.D., Besser, T.E., Rice, D.H., Ebel, E.D., Herriott, D.E. and Carpenter, L.V. Multiple sources of *Escherichia coli* O157 in feedlots and dairy farms in the northwestern USA. *Preventive Veterinary Medicine* 1998; 35: 11–19.

Jordan, D. and McEwen, S.A. Effect of duration of fasting and a short-term high-roughage ration on the concentration of *Escherichia coli* biotype 1 in cattle faeces. *Journal of Food Protection* 1998; 61: 531–4.

Kohler, B., Karch, H. and Schmidt, H. Antibacterials that are used as growth promoters in animal husbandry can affect the release of Shiga-toxin-2-converting bacteriophages and Shiga-toxin-2 from *Escherichia coli* strains. *Microbiology* 2000; 146: 1085–90.

Kudva, I.T., Blanch, K. and Hovde, C.J. Analysis of *Escherichia coli* O157:H7 survival in ovine or bovine manure and manure slurry. *Applied & Environmental Microbiology* 1998; 64: 3166–74.

Mead, G.C. Prospects for 'competitive exclusion' treatment to control salmonellas and other foodborne pathogens in poultry [see comments]. [Review] [84 refs]. *Veterinary Journal* 2000; 159: 111–23.

Mechie, S.C., Chapman, P.A. and Siddons, C.A. A fifteen month study of *Escherichia coli* O157:H7 in a dairy herd. *Epidemiology & Infection* 1997; 118: 17–25.

Paiba, G.A., Wilesmith, J.W. and Evans, S.J. Excretion of VTEC O157 by cattle [letter]. *Veterinary Record* 1999; 144: 708.

Porter, J., Mobbs, K., Hart, C.A., Saunders, J.R., Pickup, R.W. and Edwards, C. Detection, distribution and probable fate of *Escherichia coli* O157 from asymptomatic cattle on a dairy farm. *Journal of Applied Microbiology* 1997; 83: 297–306.

Rahn, K., Renwick, S.A., Johnson, R.P., Wilson, J.B., Clarke, R.C. and Alves, D. *et al.* Persistence of *Escherichia coli* O157:H7 in dairy cattle and the dairy farm environment. *Epidemiology & Infection* 1997; 119: 251–9.

Randall, L.P., Wray, C. and Davies, R.H. Survival of verocytotoxin-producing Escherichia coli O157 under simulated farm conditions. *Veterinary Record* 1999; 145: 500–1.

Russell, J.B., Diez-Gonzalez, F. and Jarvis, G.N. Potential effect of cattle diets on the transmission of pathogenic *Escherichia coli* to humans. [Review] [54 refs]. *Microbes & Infection* 2000; 2: 45–53.

Shere, J.A., Bartlett, K.J. and Kaspar, C.W. Longitudinal study of *Escherichia coli* O157:H7 dissemination on four dairy farms in Wisconsin. *Applied & Environmental Microbiology* 1998; 64: 1390–9.

CONTROL POINT 2: CONTAMINATION OF RAW MEAT

Aim:	**To prevent manure/gut contents contaminating carcasses and raw meat**

Agencies: MAFF, MHS, FSA, Veterinary Laboratory Agency, abattoir proprietors and workers, farmers' unions and representative organisations (e.g. NFU).

Legislation

The Fresh Meat (Hygiene and Inspection) Regulations 1995 (as amended).
The Meat Products (Hygiene) Regulations 1994 (as amended).

Enforcement arrangements

Hygiene in slaughterhouses is enforced by the Meat Hygiene Service. It is responsible for veterinary supervision, meat inspection and enforcement in licensed fresh meat, poultry meat and game meat establishments and in integrated premises producing meat products, meat preparations and minced meat. The MHS is audited by the Veterinary Public Health Unit of the Food Standards Agency and by the State Veterinary Service in Scotland and Wales.

Slaughterhouses must be licensed by the MHS before they can operate and licences can be revoked for failure to comply with hygiene requirements. Oral and written advice and guidance and the service of statutory notices and prosecution are further enforcement options. There are currently about 250 full throughput licensed fresh meat slaughterhouses in the UK. Details of revoked licences and a summary of enforcement action are publicly available in the Food Standards Agency publication *Meat Hygiene Enforcement Report*.

Hygiene standards in slaughterhouses are monitored and recorded by a risk-based method termed the Hygiene Assessment System (HAS). Premises are assessed against performance criteria and awarded a HAS score which can have a value between 0 and 100 with higher scores indicating better hygiene. Assessments are carried out by the Official Veterinary Surgeon at least once a month. A summary of monthly scores for all named slaughterhouses in the UK is published in the *Meat Hygiene Enforcement Report*.

The Health Mark is the key safety control for fresh meat. It must be applied to all meat sold for human consumption in the UK and indicates that the meat has been produced according to statutory hygiene standards. This is achieved through ante-mortem and post-mortem inspections. One of the purposes of ante-mortem inspection is to ensure that the animals are not contaminated by manure which could be a source of pathogenic organisms on the finished carcass.

The slaughter of dirty livestock is prohibited by the Fresh Meat (Hygiene and Inspection) Regulations 1995 as it prevents the hygienic dressing of the carcass. Five levels of fleece or hide contamination are recognised and only animals in the cleanest two categories (1 and 2) are normally accepted for slaughter. Enforcement and remedial action taken with respect of dirty animals (e.g. detention in lairage to dry or cleaning) are summarised in the *Meat Hygiene Enforcement Report*.

Published UK guidance/recommendations

ACMSF: Recommendation for research into the effectiveness of processing aids such as carcass washes in reducing the microbiological load on carcasses. Recommendation that all relevant sectors of the food industry adopt a HACCP-based approach to prevent survival of or contamination by VTEC.

Pennington: There should be an education/awareness programme for farm workers, repeated and updated periodically as appropriate to ensure that they are aware of the absolute requirement for the presentation of animals in an appropriate clean condition for slaughter.

The Meat Hygiene Service should urgently implement its scoring system for clean/dirty animals, should ensure that official veterinary surgeons and the trade are educated and trained in its use, and should pursue consistent and rigorous enforcement.

The Meat Hygiene service must take forward urgently, with the help and support of government departments and the industry, the identification and promotion of good practice in slaughterhouses – including specifically in the areas of hide and intestine removal.

Abattoir workers should be trained in good hygiene practice during slaughter and the Meat Hygiene Service should concentrate enforcement on slaughter and subsequent handling of carcasses.

The Hazard Analysis and Critical Control Point system should be enshrined in the legislation governing slaughterhouses and the transportation of carcasses and meat. Meanwhile, enforcers and the trade should ensure that HACCP principles are observed.

The Meat Hygiene Service should be given additional powers to enforce at the abattoir standards for the transportation of carcasses between licensed and non-licensed premises.

Further consideration should be given, involving the industry and consumer interests, to the potential use and benefits of end-process treatments such as steam pasteurisation.

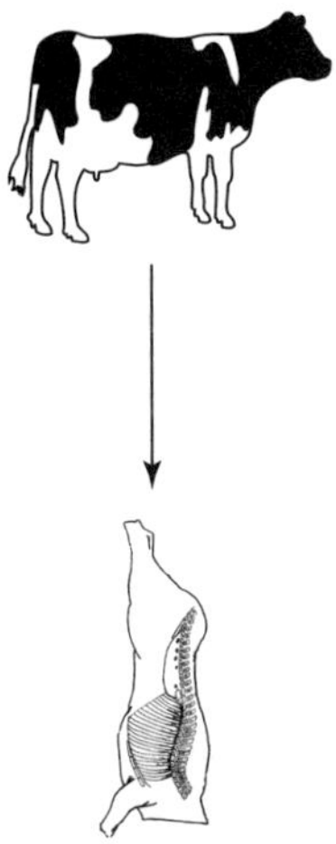

Figure 7.3 Control Point 2: Contamination of raw meat

In line with the approach recommended for more general enforcement, the efforts and resources of the Meat Hygiene Service should be targeted at higher risk premises – especially those abattoirs with Hygiene Assessment Scores (HAS score) of under 65.

Evidence presented to VTEC taskforce

Issues include cleaning, ventilation and bedding of animal transportation vehicles.

WHO: Animals for slaughter should be transported in a manner that maintains cleanliness, minimises contamination from other animals and limits undue stress.

Transport vehicles should be cleaned and dried before any further use.

Any water used for washing and/or processing of animals should be of potable quality.

Holding of animals prior to slaughter should be done in such a manner that the animals are kept clean.

All abattoirs/establishments where animals are slaughtered should have in place a system of good hygienic practice and an effective HACCP plan covering all stages in the production process, from the time the animal arrives until the carcass or meat products leave the establishment.

In handling carcasses, particular care should be taken during the dehiding stage, removal of hooves, evisceration, and cutting, to minimise cross-contamination. In addition, care should be taken to minimise spillage of intestinal tract contents.

Good abattoir practices should include the minimal use of water, especially where this contributes to the spread of faecal material.

It is advised that food producers purchasing meat from abattoirs, or those that have their own slaughter facilities, undertake regular hygiene inspections of the slaughter premises. In addition, food producers should establish raw material specifications. This has been shown to improve the quality of meat, reducing the risk of EHEC being present on raw material.

Key points and practical issues

- Accepting that some animals will be infected with VTEC O157 when they leave farms and are transported for slaughter it is likely that cross-contamination between animals, and subsequently between carcasses, will increase the overall risks to public health. VTEC O157 is present in the infected animals' manure and gut contents. Therefore, control measures prior to slaughter are aimed at preventing manure from contaminating animal hides. Washing cattle with a power hose prior to slaughter can actually increase cross-contamination by splashing hands and knives of operatives. There is currently no evidence that this method decreases the number of VTEC O157 transferred from hide to carcass during slaughter.
- During slaughter, measures to prevent spillage of gut contents and to prevent carcass-to-carcass cross-contamination are recommended. Additionally, processing of the animal carcass (including water, steam or acid spraying) may reduce bacterial contamination. Research into the efficacy of various processes is underway. Spray washing with water, 2 per cent acetic acid, acidified sodium chloride and steam has been investigated. Whilst a reduction of bacterial numbers may be achieved under experimental conditions, in practice, other factors are likely to reduce the effectiveness of such measures. VTEC O157 mixed in with animal faeces is harder to remove than when applied as a suspension. The manure adheres more strongly to the carcass surface. Operative training and equipment maintenance are also likely to play a part. Consequently, carcass treatments may reduce numbers of bacteria but are unlikely to completely remove contamination.

Further reading

Berry, E.D. and Cutter, C.N. Effects of acid adaptation of Escherichia coli O157:H7 on efficacy of acetic acid spray washes to decontaminate beef carcass tissue. *Applied & Environmental Microbiology* 2000; 66: 1493–8.

Byrne, C.M., Bolton, D.J., Sheridan, J.J., McDowell, D.A. and Blair, I.S. The effects of preslaughter washing on the reduction of Escherichia coli O157:H7 transfer from cattle hides to carcasses during slaughter. *Letters in Applied Microbiology* 2000; 30: 142–5.

Castillo, A., Lucia, L.M., Kemp, G.K. and Acuff, G.R. Reduction of Escherichia coli O157:H7 and Salmonella typhimurium on beef carcass surfaces using acidified sodium chlorite. *Journal of Food Protection* 1999; 62: 580–4.

Delazari, I., Iaria, S.T., Riemann, H., Cliver, D.O. and Jothikumar, N. Removal of Escherichia coli O157:H7 from surface tissues of beef carcasses inoculated with wet and dry manure. *Journal of Food Protection* 1998; 61: 1265–8.

CONTROL POINT 3: CONTAMINATION OF MILK

Aim: To prevent contamination of milk and dairy products

Agencies: MAFF, Dairy Hygiene Inspectorate, local authorities, farmers' unions and representative organisations (e.g. NFU), VLA, dairy industry.

Legislation

The Dairy Products (Hygiene) Regulations 1995 (as amended)

Enforcement arrangements

The regulations govern the hygienic production of dairy products. They are enforced by the Dairy Hygiene Inspectorate at production holdings and local authority environmental health departments at dairies (see below for definitions). Failure to comply with the regulations can result in revocation of registration and/or fines and imprisonment.

Published UK guidance/recommendations

ACMSF: We recommend that all relevant sectors of the food industry adopt a HACCP-based approach to prevent survival of or contamination by VTEC.

We strongly urge the government to reconsider its position concerning a ban on the sale of raw cows' milk in England, Wales and Northern Ireland. In the meantime, vulnerable groups in particular should be advised by the government's Chief Medical Officer not to consume it, and the labelling of raw cows' milk should be altered accordingly.

We recommend that industry ensures that the pasteurisation of milk and milk products is carefully controlled and that post-pasteurisation contamination is avoided.

We recommend that industry label cheese made from raw milk from cows and other species so that consumers can identify it.

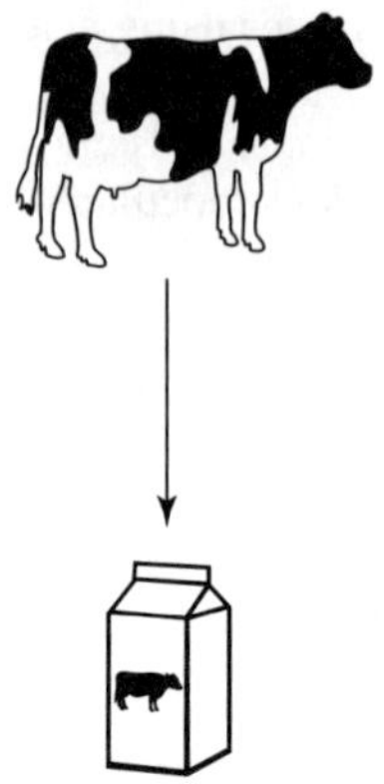

Figure 7.4 Control Point 3: Contamination of milk

We recommend that government and industry fund research and surveillance into the prevalence of VTEC O157 in raw cows' milk, cream made from raw cows' milk and raw milk cheeses.

Pennington: There should be an education/awareness programme for farm workers, repeated and updated periodically as appropriate to ensure that they are aware of the existence, potential prevalence and nature of *E.coli* O157.

Evidence presented to VTEC taskforce

No specific recommendations.

WHO: Experts agreed that it is practically impossible to produce raw milk without faecal contamination, and felt strongly that to prevent EHEC transmission from milk, it should be pasteurised or treated with procedures able to deliver an equal level of safety.

Post-pasteurisation contamination in the plant or after packaging must be avoided by adherence to a comprehensive HACCP plan.

Manufacturers of fermented foods (such as yoghurt or cheese made from unpasteurised milk) which use raw materials that may harbour EHEC should assure their absence in a serving portion (demonstrate that their process is sufficient to remove or destroy the organisms in a serving portion).

Key points and practical issues

- Raw milk can be contaminated by animal manure during the milking process and it may also be possible for VTEC O157 to be excreted in the milk. Several outbreaks have been attributed to unpasteurised milk, cream and cheese.
- The sale of raw cows' milk has been banned in Scotland since 1983 and in England and Wales it is restricted (by the above regulations) to sales made directly to consumers on the farm, farm gate or milk round. Following a public consultation between November 1997 and February 1998, a decision was taken not to ban the sale of unpasteurised milk in England and Wales. Instead the frequency of official microbiological testing at production premises was increased to four times a year and the frequency of official inspections of registered raw cows' milk production holdings was increased to once yearly with immediate effect (1999). It is a requirement that raw cows' milk is labelled with a public health warning which must include the Chief Medical Officer's advice (issued 1995) that pregnant women, elderly people and those currently unwell or who have a chronic illness should not consume raw cows' milk.
- Raw milk and dairy products are also a source of cross-contamination for other food products including pasteurised milk and dairy products. Research suggests that VTEC O157 survives well even in fermented dairy products such as yoghurt, sour cream and buttermilk. Several outbreaks have been caused by post-pasteurisation contamination or labelling errors. Poor temperature control after contamination adds to the public health risk.
- Unpasteurised goat and sheep milk are also potential source of human infection.
- The Dairy Products (Hygiene) Regulations 1995 apply to all milk intended for human consumption produced from cows, sheep, goats and buffaloes. The premises where the milk is produced are referred to as 'production holdings'. Premises where milk undergoes heat treatment or processing (pasteurisation, manufacture, bottling) are termed 'dairy establishments'. Production holdings have to attain minimum standards of design, situation, construction and fitting, separation of animals from milk storage areas and temperature control of milk.
- Dairy establishments can only use milk produced by registered production holdings or other approved establishments. The structure and layout of the premises must render them easy to clean and disinfect. Microbiological standards for raw milk intended for heat treatment and drinking milk are laid down together with temperature requirements. Adequate heat treatment of raw milk destroys VTEC O157.

Further reading

Ansay, S.E. and Kaspar, C.W. Survey of retail cheeses, dairy processing environments and raw milk for Escherichia coli O157:H7. *Letters in Applied Microbiology* 1997; 25: 131–4.

Bielaszewska, M., Janda, J., Blahova, K., Minarikova, H., Jikova, E. and Karmali, M.A. *et al.* Human Escherichia coli O157:H7 infection associated with the consumption of unpasteurised goat's milk. *Epidemiology & Infection* 1997; 119: 299–305.

Cases of Escherichia coli O157 infection associated with unpasteurised cream. *Communicable Disease Report* 1998; *CDR Weekly* 8: 377.

Clark, A., Morton, S., Wright, P., Corkish, J., Bolton, F.J. and Russell J. A community outbreak of Vero cytotoxin producing Escherichia coli O157 infection linked to a small farm dairy. *Communicable Disease Report* 1997; *CDR Review* 7: R206–R211.

Dineen, S.S., Takeuchi, K., Soudah, J.E. and Boor, K.J. Persistence of Escherichia coli O157:H7 in dairy fermentation systems. *Journal of Food Protection* 1998; 61: 1602–8.

Escherichia coli O157 in Somerset. *Communicable Disease Report* 1998; *CDR Weekly* 8: 167.

Escherichia coli O157 associated with eating unpasteurised cheese: update. *Communicable Disease Report* 134; *CDR Weekly* 9: 131.

Escherichia coli O157 associated with eating unpasteurised cheese. *Communicable Disease Report* 116; *CDR Weekly* 9: 113.

Guraya, R., Frank, J.F. and Hassan, A.N. Effectiveness of salt, pH, and diacetyl as inhibitors for Escherichia coli O157:H7 in dairy foods stored at refrigeration temperatures. *Journal of Food Protection* 1998; 61: 1098–102.

Heuvelink, A.E., Bleumink, B., van den Biggelaar, F.L., Te Giffel, M.C., Beumer, R.R. and de Boer, E. Occurrence and survival of verocytotoxin-producing Escherichia coli O157 in raw cow's milk in The Netherlands. *Journal of Food Protection* 1998; 61: 1597–601.

Ingham, S.C., Su, Y.C. and Spangenberg, D.S. Survival of Salmonella typhimurium and Escherichia coli O157:H7 in cheese brines. *International Journal of Food Microbiology* 2000; 61: 73–9.

Issa, M.S. and Ryser, E.T. Fate of Listeria monocytogenes, Salmonella typhimurium DT104, and Escherichia coli O157:H7 in Labneh as a pre- and postfermentation contaminant. *Journal of Food Protection* 2000; 63: 608–12.

Keene, W.E., Hedberg, K., Herriott, D.E., Hancock, D.D., McKay, R.W. and Barrett, T.J. *et al.* A prolonged outbreak of Escherichia coli O157:H7 infections caused by commercially distributed raw milk. *Journal of Infectious Diseases* 1997; 176: 815–18.

Massa, S., Altieri, C., Quaranta, V. and De Pace, R. Survival of Escherichia coli O157:H7 in yoghurt during preparation and storage at 4 degrees C. *Letters in Applied Microbiology* 1997; 24: 347–50.

Massa, S., Goffredo, E., Altieri, C. and Natola, K. Fate of Escherichia coli O157:H7 in unpasteurised milk stored at 8 degrees C. *Letters in Applied Microbiology* 1999; 28: 89–92.

McIngvale, S.C., Chen, X.Q., McKillip, J.L. and Drake, M.A. Survival of Escherichia coli O157:H7 in buttermilk as affected by contamination point and storage temperature. *Journal of Food Protection* 2000; 63: 441–4.

Mechie, S.C., Chapman, P.A. and Siddons, C.A. A fifteen month study of Escherichia coli O157:H7 in a dairy herd. *Epidemiology & Infection* 1997; 118: 17–25.

Outbreak of Vero cytotoxin producing Escherichia coli O157 infection in north Cumbria. *Communicable Disease Report* 1998; *CDR Weekly* 9: 95.

Outbreaks of VTEC O157 infection linked to consumption of unpasteurised milk. *Communicable Disease Report* 206; *CDR Weekly* 10: 203.

Ramsaran, H., Chen, J., Brunke, B., Hill, A. and Griffiths, M.W. Survival of bioluminescent Listeria monocytogenes and Escherichia coli O157:H7 in soft cheeses. *Journal of Dairy Science* 1998; 81: 1810–17.

Rubini, S., Cardeti, G., Amiti, S., Manna, G., Onorati, R. and Caprioli, A. *et al.* Verocytotoxin-producing Escherichia coli O157 in sheep milk [letter]. *Veterinary Record* 1999; 144: 56.

Vero cytotoxin producing Escherichia coli O157. *Communicable Disease Report* 412; *CDR Weekly* 7: 409.

VTEC O157 infection and unpasteurised cream: update. *Communicable Disease Report* 392; *CDR Weekly* 8: 389.

VTEC O157 phage type 21/28 infection in North Cumbria: update. *Communicable Disease Report* 1999; *CDR Weekly* 9: 105.

CONTROL POINT 4: CONTAMINATION OF CROPS AND PRODUCE (FRUIT AND VEGETABLES) BY ANIMAL MANURE

Aim: To prevent manure/slurry containing VTEC O157 from contaminating fresh fruit and vegetables

Agencies: MAFF, Veterinary Laboratory Agency, farmers' unions and representative organisations (e.g. NFU), FSA, local authorities.

Legislation

Environmental Protection Act 1990
Water Resources Act 1991
Control of Pollution (Silage, Slurry and Agricultural Fuel Oil) Regulations 1991 (as amended).

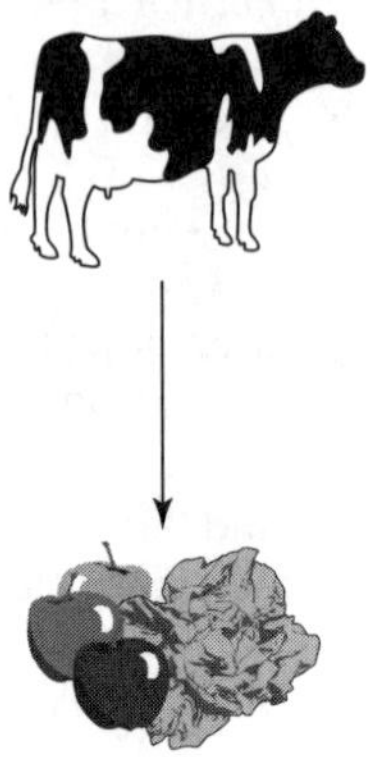

Figure 7.5 Control Point 4: Contamination of crops and produce by animal manure

Enforcement arrangements

The Environment Agency monitors and enforces the provisions of the above legislation. Several codes of practice and guidance leaflets have been compiled relating to the storage and use of animal slurry on agricultural land. Farm waste management plans are encouraged to carefully plan the most appropriate methods and timing of slurry application.

Published UK guidance/recommendations

ACMSF: We recommend that all relevant sectors of the food industry adopt a HACCP-based approach to prevent survival of or contamination by VTEC.

Pennington: There should be an education/awareness programme for farm workers, repeated and updated periodically as appropriate to ensure that they are aware of the existence, potential prevalence and nature of *E.coli* O157, the potential for the spread of infection on farms in a number of ways, including notably from faecal material and of the need for care in the use of untreated slurry or manure.

Evidence presented to VTEC taskforce: Subjects covered included: clarification of standards for composting, manure storage and treatment, research/education on effect of manure application, post-harvest washing and risks from organic food and care in the use of natural surface waters (especially those which receive run-off from agricultural land) for irrigation of 'ready-to-eat' produce, e.g. salad crops.

WHO: Animal slurry should not be used on or near crops intended for human consumption, unless it has been adequately treated. WHO guidelines for the safe use of wastewater and excreta in agriculture and aquaculture recommend that treated wastewater should contain <1,000 faecal coliforms per 100 ml for unrestricted irrigation in agriculture, and <10,000 faecal coliforms per 100 ml/g for unrestricted use in aquaculture.

Crops used for raising seeds that are going to be used as sprouts should not be fertilised with animal slurry or human faecal waste, even if some form of treatment has been applied.

Crops for food should be irrigated with water that is not faecally contaminated. Any water used for washing and/or processing of fruits and vegetables should be of potable quality.

Special care should be paid to fruit and other produce to prevent the accidental contact with soil or animal faeces (e.g. fallen apples) as these have been implicated in the transmission of EHEC infections.

For raw, ready-to-drink fruit or vegetable juices where safety of the product cannot be assured, a processing step which can remove or destroy EHEC, such as pasteurisation, should be applied.

Irradiation should be considered as a decontamination step for some products, especially if there is no other practical control or prevention step available, or if they are ready-to-eat foods for highly susceptible individuals.

Manufacturers of fermented foods (such as cereal products) which use raw materials that may harbour EHEC should assure their absence in a serving portion (demonstrate that their process is sufficient to remove or destroy the organisms in a serving portion).

Producers of sprouts from seeds (manufacturers, retailers, food service establishments and home-makers) should use potable water.

Key points and practical issues

- Applying animal manure to land used for growing crops (fruit and vegetables) for human consumption or allowing raw fruit and vegetables (e.g. apples) to land on pasture grazed by animals provides a route of transmission for VTEC O157 into the food chain. Several outbreaks have been reported worldwide. In particular, apples (and unpasteurised apple juice), lettuce and sprouting seeds (alfafa, bean sprouts, mung beans) have been implicated. The largest outbreak involving white radish sprouts from a single farm was recorded in Japan in July 1996. It has been demonstrated that VTEC O157 can adhere (particularily to cut surfaces) to lettuce. Manure does not necessarily have to be applied directly to the crop. Cattle grazing near lettuce fields were thought to be the source of contamination for an outbreak involving mesclun lettuce. Wind-blown apples landing on pasture grazed by cattle were the source of infection in a further outbreak.

- A comprehensive approach to good manufacturing practice is needed to protect the food from initial contamination. Research into further control measures is currently being undertaken. Ionising radiation and washing with various chemicals (chlorine, allyl isothiocyanate) are potential methods of disinfecting such produce although simply washing with chlorinated water appears to have little effect. Current chemical treatments used to eliminate pathogens from sprouting seeds are not necessarily effective against VTEC O157. Viable VTEC O157 have been isolated from radish sprouts. Finally, adequate heat treatment of produce prior to consumption may be feasible in some cases (some sprouting seeds) but is clearly impossible in others (e.g. lettuce).

Further reading

Ackers, M.L., Mahon, B.E., Leahy, E., Goode, B., Damrow, T. and Hayes, P.S. *et al.* An outbreak of Escherichia coli O157:H7 infections associated with leaf lettuce consumption. *Journal of Infectious Diseases* 1998; 177: 1588–93.

Beuchat, L.R. Survival of enterohemorrhagic Escherichia coli O157:H7 in bovine feces applied to lettuce and the effectiveness of chlorinated water as a disinfectant. *Journal of Food Protection* 1999; 62: 845–9.

Beuchat, L.R., Nail, B.V., Adler, B.B. and Clavero, M.R. Efficacy of spray application of chlorinated water in killing pathogenic bacteria on raw apples, tomatoes, and lettuce. *Journal of Food Protection* 1998; 61: 1305–11.

Delaquis, P.J., Sholberg, P.L. and Stanich, K. Disinfection of mung bean seed with gaseous acetic acid. *Journal of Food Protection* 1999; 62: 953–7.

From the Centers for Disease Control and Prevention. Outbreaks of Escherichia coli O157:H7 infection associated with eating alfalfa sprouts: Michigan and Virginia, June–July 1997. *JAMA* 1997; 278: 809–10.

Hilborn, E.D., Mermin, J.H., Mshar, P.A., Hadler, J.L., Voetsch, A. and Wojtkunski, C. *et al.* A multistate outbreak of Escherichia coli O157:H7 infections associated with consumption of mesclun lettuce. *Archives of Internal Medicine* 1999; 159: 1758–64.

Itoh, Y., Sugita-Konishi, Y., Kasuga, F., Iwaki, M., Hara-Kudo, Y. and Saito, N. *et al.* Enterohemorrhagic Escherichia coli O157:H7 present in radish sprouts. *Applied & Environmental Microbiology* 1998; 64: 1532–5.

Lang, M.M., Ingham, B.H. and Ingham, S.C. Efficacy of novel organic acid and hypochlorite treatments for eliminating Escherichia coli O157:H7 from alfalfa seeds prior to sprouting. *International Journal of Food Microbiology* 2000; 58: 73–82.

Lin, C.M., Kim, J., Du, W.X. and Wei, C.I. Bactericidal activity of isothiocyanate against pathogens on fresh produce. *Journal of Food Protection* 2000; 63: 25–30.

Michino, H., Araki, K., Minami, S., Takaya, S., Sakai, N. and Miyazaki, M. *et al.* Massive outbreak of Escherichia coli O157:H7 infection in schoolchildren in Sakai City, Japan, associated with consumption of white radish sprouts [see comments]. *American Journal of Epidemiology* 1999; 150: 787–96.

Park, C.M., Taormina, P.J. and Beuchat, L.R. Efficacy of allyl isothiocyanate in killing enterohemorrhagic Escherichia coli O157:H7 on alfalfa seeds. *International Journal of Food Microbiology* 2000; 56: 13–20.

Rajkowski, K.T. and Thayer, D.W. Reduction of Salmonella spp. and strains of Escherichia coli O157:H7 by gamma radiation of inoculated sprouts. *Journal of Food Protection* 2000; 63: 871–5.

Seo, K.H. and Frank, J.F. Attachment of Escherichia coli O157:H7 to lettuce leaf surface and bacterial viability in response to chlorine treatment as demonstrated by using confocal scanning laser microscopy. *Journal of Food Protection* 1999; 62: 3–9.

Takeuchi, K. and Frank, J.F. Penetration of Escherichia coli O157:H7 into lettuce tissues as affected by inoculum size and temperature and the effect of chlorine treatment on cell viability. *Journal of Food Protection* 2000; 63: 434–40.

Taormina, P.J. and Beuchat, L.R. Behavior of enterohemorrhagic Escherichia coli O157:H7 on alfalfa sprouts during the sprouting process as influenced by treatments with various chemicals. *Journal of Food Protection* 1999; 62: 850–6.

Taormina, P.J. and Beuchat, L.R. Comparison of chemical treatments to eliminate enterohemorrhagic Escherichia coli O157:H7 on alfalfa seeds. *Journal of Food Protection* 1999; 62: 318–24.

Taormina, P.J., Beuchat, L.R. and Slutsker, L. Infections associated with eating seed sprouts: an international concern. [Review] [47 refs]. *Emerging Infectious Diseases* 1999; 5: 626–34.

Watanabe, Y., Ozasa, K., Mermin, J.H., Griffin, P.M., Masuda, K. and Imashuku, S. *et al.* Factory outbreak of Escherichia coli O157:H7 infection in Japan. *Emerging Infectious Diseases* 1999; 5: 424–8.

CONTROL POINT 5: CONTAMINATION OF CROPS AND PRODUCE (FRUIT AND VEGETABLES) BY HUMAN FAECES

Aim: To prevent human sewage sludge from contaminating fresh fruit and vegetables

Agencies: MAFF, The Environment Agency, farmers' unions and representative organisations (e.g. NFU), FSA.

Legislation

Environmental Protection Act 1990.
Waste Management Licensing Regulations 1994.
Sludge (Use in Agriculture) Regulations 1989.

Enforcement arrangements

The 1994 regulations apply to the disposal of non-agricultural waste by spreading on land. Such activities are licensed and monitored by the Environment Agency. Small amounts (no more than 250 tonnes per hectare in any 12 month period) of blood and gut contents from abattoirs and septic tank sludge, both potential sources of VTEC O157, can be applied to agricultural land without licence so long as certain conditions are met and that the Environment Agency is notified.

More than half of the sewage sludge produced in England and Wales is applied to agricultural land as organic fertiliser. Further quantities are used in forestry and land restoration. The Environment Agency monitors the provisions of the regulations which include measures to protect crops intended for human consumption. The harvesting of forage crops or the grazing of animals on agricultural land to which treated sludge has been surface applied is prohibited for at least three weeks after application. Fruit or other crops normally eaten raw must not be harvested for at least 10 months after sewage sludge has been applied.

The 1989 regulations are due to be amended to include the Agricultural Development Advisory Service safe sludge matrix. As from 31 December 1999, untreated sludges cannot be applied to food crops. As of 31 December 1998, conventionally treated sewage sludge can only be applied to grazed land if it is deep injected into the soil and if there is no grazing or harvesting within 3 weeks. Conventionally treated sludge (usually by anaerobic digestion which kills 99 per cent of contaminating pathogens) can be applied to land used to grow vegetables (including potatoes, parsnips and leeks) at least 12 months before a crop is harvested. Where the crop might be eaten raw (lettuce, onions, carrots, cabbage, etc.) the harvest interval must be at least 30 months. Where enhanced treated sludges are used (capable of producing a 6 log reduction in pathogens) a 10 month harvest interval is required.

Published UK guidance/recommendations

ACMSF: We recommend that all relevant sectors of the food industry adopt a HACCP-based approach to prevent survival of or contamination by VTEC.

Pennington: There should be an education/awareness programme for farm workers, repeated and updated periodically as appropriate to ensure that they are aware of the existence, potential prevalence and nature of *E.coli* O157, the potential for the spread of infection on farms in a number of ways, including notably from faecal material and of the need for care in the use of untreated slurry or manure.

Evidence presented to VTEC taskforce: More control of imported off-farm wastes (cf. sewage sludge); sewage sledges to be treated in accordance with ADAS matrix and tested prior to recycling to land; all organic and beneficial materials including blood and blood products to be treated to a recognised and robust set of standards prior to recycling; consider increasing 3 weeks no grazing after slurry application to 4 weeks; follow appropriate guidelines for application of slurry. Adherence to 'Safe Sludge Matrix'.

WHO: Human faecal waste should not be used on or near crops intended for human consumption, unless it has been adequately treated. WHO guidelines for the safe use of wastewater and excreta in agriculture and aquaculture recommend that treated wastewater should contain <1,000 faecal coliforms per 100 ml for unrestricted irrigation in agriculture, and <10,000 faecal coliforms per 100 ml/g for unrestricted use in aquaculture.

Crops used for raising seeds that are going to be used as sprouts should not be fertilised with human faecal waste, even if some form of treatment has been applied.

Crops for food should be irrigated with water that is not faecally contaminated. Any water used for washing and/or processing of fruits and vegetables should be of potable quality.

Irradiation should be considered as a decontamination step for some products, especially if there is no other practical control or prevention step available, or if they are ready-to-eat foods for highly susceptible individuals.

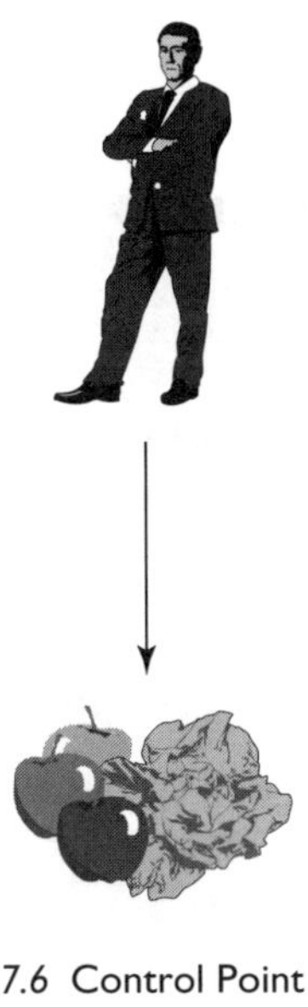

Figure 7.6 Control Point 5: Contamination of crops and produce by human faeces

Manufacturers of fermented foods (such as cereal products) which use raw materials that may harbour EHEC should assure their absence in a serving portion (demonstrate that their process is sufficient to remove or destroy the organisms in a serving portion).

Producers of sprouts from seeds (manufacturers, retailers, food service establishments and home-makers) should use potable water.

Key points and practical issues

- Sewage sludge arises from the treatment of human sewage. Untreated sludge is obtained from the settlement or biological stages of treatment. Further processing can be undertaken to improve its stability. It is applied to land as semi-liquid (similar to animal slurry) or as de-watered sludge cake.
- Sludge from septic tanks and cesspools are also commonly spread on agricultural land. Septic tank waste is partially digested sewage and can be spread on agricultural land with the same controls as applied to sewage sludge. Cesspool waste is raw and undigested. Spreading cesspool waste must be licensed by the Environment Agency under the 1994 regulations.

Further reading

Gagliardi, J.V. and Karns, J.S. Leaching of Escherichia coli O157:H7 in diverse soils under various agricultural management practices. *Applied & Environmental Microbiology* 2000; 66: 877–83.

CONTROL POINT 6: CONTAMINATION OF DRINKING WATER BY ANIMAL MANURE

Aim: To prevent manure/slurry containing VTEC O157 from contaminating drinking water supplies

Agencies: MAFF, The Environment Agency, local authority environmental health departments, Drinking Water Inspectorate, farmers' unions and representative organisations (e.g. NFU), FSA.

Legislation

Environmental Protection Act 1990.

Water Resources Act 1991.
Control of Pollution (Silage, Slurry and Agricultural Fuel Oil) Regulations 1991 (as amended).
Water Industry Act 1991.
The Private Water Supplies Regulations 1991.
Natural Mineral Water Regulations 1985.
The Water Supply (Water Quality) Regulations 1989 (as amended).
The Surface Waters (Abstraction for Drinking Water) (Classification) Regulations 1996.

Enforcement arrangements

The microbiological quality of surface water intended for the abstraction of drinking water, for bathing water, freshwater fish, shellfish and for water intended for human consumption is governed by European directives, such as that governing surface water (Directive 75/440/EEC), bathing water (Directive 76/160/EEC) and water intended for human consumption (Directive 80/778). The directives are implemented through legislation.

The Environment Agency monitors and controls pollution of groundwater and watercourses. Drinking water is supplied in England and Wales by private companies, in Scotland by publicly owned water authorities and in Northern Ireland by the Northern Ireland Water Service. Section 70 of the Water Industry Act 1991 requires the water companies to supply water which is fit for human consumption.

In England and Wales, the performance of water companies in providing wholesome water complying with the Water Supply (Water Quality) Regulations is monitored by the Drinking Water Inspectorate. The DWI carries out technical audits of water companies which include assessments of sampling and progress with improvement programmes. Where companies are in breach of the regulations, the DWI issues a notification that enforcement action is being considered. If no evidence is provided that remedial action to prevent a recurrence has been taken, an undertaking to carry out works by a specified date is accepted from the company or an enforcement order is served.

The DWI also investigates incidents where water companies have supplied water that was unfit for human consumption and can bring prosecutions for this offence. Summaries of improvement programmes and enforcement action together with incidents and prosecutions are publicly available on the DWI website.

Sources of contamination for water supplies from farmed animals are manure, slurry (manure mixed with water), and dirty water (from milking parlours, farm dairies, run-off from yards, etc.)

The Water Resources Act 1991 prohibits the known discharge of poisonous, noxious or polluting matter or solid waste into any 'controlled water'.

'Controlled waters' include groundwaters and all inland waters including lakes, ponds, rivers, streams, canals and field ditches. The legislation is enforced by the Environment Agency and fines and imprisonment can be imposed on farmers, employees or contractors as well as costs for prosecution and remedial work.

The Control of Pollution (Silage, Slurry and Agricultural Fuel Oil) Regulations 1991 (as amended) prevent pollution by slurry and dirty water by setting standards for storing and handling these substances on farms. Detailed guidance on the design and operation of slurry storage facilities is available in 'The Water Code' (MAFF 1998) and British Standard (BS) 5502: Part 50: 1989. The basic requirements are that slurry can only be stored in a reception pit or slurry storage tank which must be located at least 10 metres away from a watercourse or field drain and that the size and construction must prevent liquid escaping. A Farm Waste Management plan is also recommended to carefully decide the most appropriate location, time and method for spreading manure, slurry and dirty water to reduce the risk of contaminating water. Spreading within 10 metres of a ditch or water course and within 50 metres of a water source such as a spring, well or borehole should be avoided.

The amount of dirty water should be minimised by directing clean water away from sources of contamination. Dirty water should never be allowed to flow into a watercourse.

The Waste Management Licensing Regulations 1994 apply to the disposal of non-agricultural waste by spreading on land. Such activities are licensed and monitored by the Environment Agency. Small amounts (no more than 250 tonnes per hectare in any 12 month period) of blood and gut contents from abattoirs can be applied to agricultural land without licence so long as certain conditions are met and that the Environment Agency is notified.

Local authorities (in practice the Environmental Health Department) are required, by the Water Industry Act 1991, to keep themselves informed about the wholesomeness of public and private water supplies in their area. In the case of private water supplies supplied for domestic purposes or for food production to staff canteens, hostels, etc. local authorities have the responsibility to ensure that they are wholesome and can serve notices under the Water Industry Act requiring improvements. The local authorities duties with regard to safeguarding the quality of private water supplies are specified by the Private Water Supplies Regulations 1991, and with regard to the quality of bottled mineral water by the Natural Mineral Water Regulations 1985. The regulations prescribe standards for judging the wholesomeness of private water supplies including a standard of no coliforms, faecal coliforms or faecal streptococci per 100 ml. Local authorities have a duty to take samples from all private supplies serving premises in their area. The sampling frequency is determined by the use

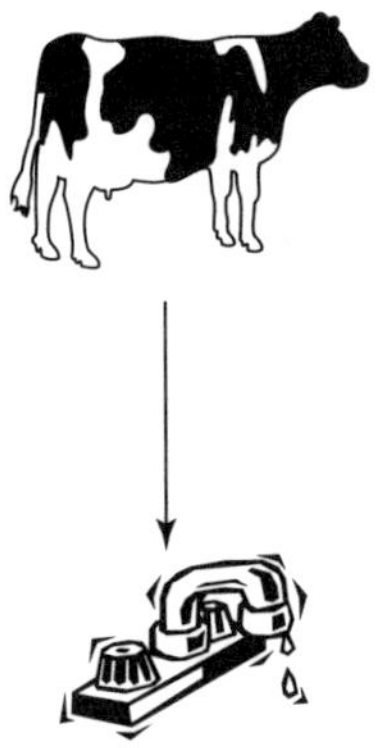

Figure 7.7 Control Point 6: Contamination of drinking water by animal manure

of the supply (domestic only or hospital, canteen, etc.) and the number of persons supplied and average daily volume of water supplied.

Where the microbiological quality of drinking water presents a danger to public health, local authorities and water companies can issue 'Boil Water Notices'.

Published UK guidance/recommendations

ACMSF: No specific recommendations.

Pennington: There should be an education/awareness programme for farm workers, repeated and updated periodically as appropriate to ensure that they are aware of the existence, potential prevalence and nature of *E.coli* O157, the potential for the spread of infection on farms in a number of ways, including notably from faecal material, and of the need for care in the use of untreated slurry or manure.

Evidence presented to VTEC taskforce: Subjects covered included: clarification of standards for composting, manure storage and treatment, proactively promote Control of Pollution Regulations, Prevention of Environmental Pollution from Agricultural Activity Code and Farm Waste Management Plans; initiate awareness raising campaign on waste storage and application for farmers and contractors, change perceptions; economic support needed for farmers; encourage waste management training for farmers and contractors; minimise volumes and leakage of contaminated water.

Continuous refinement of methodology, consideration of risks to public supplies when water treatment process or chlorination failures arise or when treated water is potentially contaminated; communication of *E.coli* and enterococci detections to relevant stakeholders; upgrade/replace poor supplies related to degree of risk and improve raw water source protection; research required into survival of O157 in water and removal by water treatment process, urgent review of the private water supplies regulations (e.g. sampling frequencies and treatment standards); compulsory treatment of private water supplies for transient and tourist populations; national protocol for action when O157 detected in private or public supplies; standardisation of boil notices/advice; further research into human immunity/susceptibility to O157; education of users on risks associated with faecal contamination of supplies; financial assistance with provision of suitable water treatment systems; encourage solicitors/mortgage providers to include water supplies as part of property survey; increase availability of mains water supply; include water supplies in building regulations; find out about your water supply and responsibility for upkeep; keep private water supply safe by protection from animals, run-off, and other possible sources of contamination; if there is suspicion that private water supply is unsafe, then water should be boiled; install treatment facilities, use bottled water, or connect to mains supply; consider other issues, e.g. lead, chemicals and microorganisms when treating a private supply; when on holiday, check whether water is from a private supply, if so, boil water for drinking and food production purposes or use bottled water; application of the microbiological risk assessment protocol to all private water supplies; provision of co-ordinated education/support for owners of private water supplies to aid them in the maintenance of such supplies; continued development of the waterborne disease surveillance programme within SCIEH; continued co-operation between SCIEH and PHLS through strengthened links between these two organisations; private water supplies should be tested regularly for faecal coliforms, especially those in areas where cattle graze or slurry or manure is spread. Attempt to limit access of wild animals/birds to human water supplies, especially untreated supplies.

Interim guidance on recreational use of animal pasture. Ensure that water from burns and streams is treated before drinking.

WHO: Animal waste must be disposed of in a manner that does not contaminate the water supply. Both run-off water and proximity of animals to wells have been shown to contaminate water supplies.

Regulatory authorities should be made aware that EHEC infections have been attributed to drinking untreated well-water contaminated by a variety of sources; and they should take appropriate action.

Key points and practical issues

- About 18,000 million litres of drinking water are supplied daily in the UK via 397,401 km of water mains. Drinking water is drawn from 1,584 boreholes, 666 reservoirs and 602 river abstractions. Two-thirds of our drinking water comes from surface water, the remainder from groundwater. There are around 2,250 treatment plants where water undergoes a series of processes which may include screening, coagulation, clarification, filtration, pH correction and a final disinfection with chlorine. The majority of drinking water in England and Wales is supplied by private water companies. About 1 per cent of the population of England and Wales have a private water supply, not supplied by a water company, which are generally located in rural areas and which serve only one or a limited number of properties. The source of the supply is usually a spring, well, borehole, stream, river, lake or pond. In general, springs, boreholes and wells are less likely to become contaminated as they draw water from deep underground although contamination can still occur at points of emergence or collection. Where water is abstracted from lakes, rivers, streams or ponds, particularly where they run through agricultural land, there is a considerable risk of contamination by VTEC O157. It has been shown that VTEC O157 can survive for at least 12 weeks in water. Microbial contamination of such supplies is particularly likely during periods of high rainfall (more run-off) and warm weather (low supplies). Various methods of disinfecting private water supplies including ultraviolet disinfection units are available. Research has shown that it is necessary to apply a UV fluence of at least 400 J/m2 to destroy VTEC O157. Electrolysed oxidizing has also been suggested as an alternative disinfection method.
- Several outbreaks, both in the UK and abroad, have been attributed to contamination of water supplies by animal manure. Untreated (private) supplies are particularly vulnerable and found mainly in agricultural areas. However, public chlorinated supplies can also be the source of outbreaks where the integrity of the supply network is breached, for example, during repairs or breakages. Waterborne outbreaks can be difficult to investigate. Contamination of the supply can be intermittent and bacteria levels in water can be very low making it almost impossible to isolate the causative organism. In one outbreak, bacteriological tests for *E.coli* were negative whereas VTEC O157 was isolated using Sorbitol MacConkey Agar. Specific methods for isolating VTEC O157 from water can be used. Alternatively, manure samples from animals suspected of contaminating the supply can be obtained. Strains can be compared using techniques such as pulsed filed gel electrophoresis. Careful epidemiological investigation of cases by gathering information on sources of water supplies and water consumption histories is useful.

Case control studies of sporadic cases have also identified consumption of unchlorinated well water as a risk factor.

Further reading

Jackson, S.G., Goodbrand, R.B., Johnson, R.P., Odorico, V.G., Alves, D. and Rahn, K. *et al.* Escherichia coli O157:H7 diarrhoea associated with well water and infected cattle on an Ontario farm. *Epidemiology & Infection* 1998; 120: 17–20.

Kurokawa, K., Tani, K., Ogawa, M. and Nasu, M. Abundance and distribution of bacteria carrying sltII gene in natural river water. *Letters in Applied Microbiology* 1999; 28: 405–10.

Outbreak of Escherichia coli O157:H7 and Campylobacter among attendees of the Washington County Fair – New York, 1999. *MMWR – Morbidity & Mortality Weekly Report* 1999; 48: 803–5.

Slutsker, L., Ries, A.A., Maloney, K., Wells, J.G., Greene, K.D. and Griffin, P.M. A nationwide case-control study of Escherichia coli O157:H7 infection in the United States. *Journal of Infectious Diseases* 1998; 177: 962–6.

Sommer, R., Lhotsky, M., Haider, T. and Cabaj, A. UV inactivation, liquid-holding recovery, and photoreactivation of Escherichia coli O157 and other pathogenic Escherichia coli strains in water. *Journal of Food Protection* 2000; 63: 1015–20.

Tanaka, Y., Yamaguchi, N. and Nasu, M. Viability of Escherichia coli O157:H7 in natural river water determined by the use of flow cytometry. *Journal of Applied Microbiology* 2000; 88: 228–36.

Venkitanarayanan, K.S., Ezeike, G.O., Hung, Y.C. and Doyle, M.P. Efficacy of electrolyzed oxidizing water for inactivating Escherichia coli O157:H7, Salmonella enteritidis, and Listeria monocytogenes. *Applied & Environmental Microbiology* 1999; 65: 4276–9.

Wang, G. and Doyle, M.P. Survival of enterohemorrhagic Escherichia coli O157:H7 in water. *Journal of Food Protection* 1998; 61: 662–7.

CONTROL POINT 7: CONTAMINATION OF RECREATIONAL WATER BY ANIMAL MANURE

Aim: To prevent manure/slurry containing VTEC O157 from contaminating recreational water

Agencies: MAFF, the Environment Agency, local authority environmental health departments, farmers unions' and representative organisations (e.g. NFU).

Legislation

Environmental Protection Act 1990.
Water Resources Act 1991.
Control of Pollution (Silage, Slurry and Agricultural Fuel Oil) Regulations 1991 (as amended).
Water Industry Act 1991.

Enforcement arrangements

The Environment Agency monitors and controls pollution of groundwater and watercourses.

Sources of contamination for water supplies from farmed animals are manure, slurry (manure mixed with water) and dirty water (from milking parlours, farm dairies, run-off from yards, etc.).

The Water Resources Act 1991 prohibits the known discharge of poisonous, noxious or polluting matter or solid waste into any 'controlled water'. 'Controlled waters' include groundwaters and all inland waters including lakes, ponds, rivers, streams, canals and field ditches. The legislation is enforced by the Environment Agency and fines and imprisonment can be imposed on farmers, employees or contractors as well as costs for prosecution and remedial work.

The Control of Pollution (Silage, Slurry and Agricultural Fuel Oil) Regulations 1991 (as amended) prevent pollution by slurry and dirty water by setting standards for storing and handling these substances on farms. Detailed guidance on the design and operation of slurry storage facilities is available in 'The Water Code' (MAFF 1998) and British Standard (BS) 5502: Part 50: 1989. The basic requirements are that slurry can only be stored in a reception pit or slurry storage tank which must be located at least 10 metres away from a watercourse or field drain and that the size and construction must prevent liquid escaping. A Farm Waste Management plan is also recommended to carefully decide the most appropriate location, time and method for spreading manure, slurry and dirty water to reduce the risk of contaminating water. Spreading within 10 metres of a ditch or watercourse should be avoided.

The amount of dirty water should be minimised by directing clean water away from sources of contamination. Dirty water should never be allowed to flow into a watercourse.

The Waste Management Licensing Regulations 1994 apply to the disposal of non-agricultural waste by spreading on land. Such activities are licensed and monitored by the Environment Agency. Small amounts (no more than 250 tonnes per hectare in any 12 month period) of blood and gut contents from abattoirs can be applied to agricultural land without licence so long as certain conditions are met and that the Environment Agency is notified.

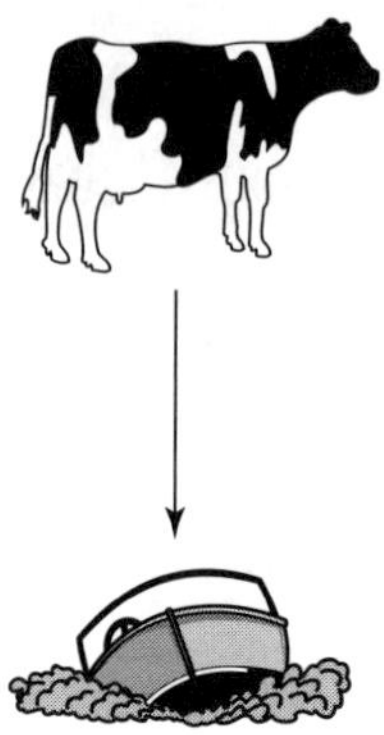

Figure 7.8 Control Point 7: Contamination of recreational water by animal manure

The microbiological quality of surface water intended for bathing is governed by European Directive 76/160/EEC. The directives are implemented through UK legislation. Directive 76/160/EEC requires member states to identify relevant bathing waters. There are currently 11 inland waters and 545 coastal waters identified as bathing waters.

The annual assessment of compliance with the Bathing Waters Directive is made by the Department of the Environment, Transport and the Regions (DETR). In practice monitoring is completed by the Environment Agency by taking samples of water. The bathing season in England and Wales runs from 15 May to 30 September and sampling commences two weeks before the start of the season. Twenty samples are taken at regular intervals throughout the season at each site. Samples are taken at points where the average density of bathers is at its highest. There must be no more than 10,000 total coliforms per 100 ml and no more than 2,000 faecal coliforms per 100 ml. In order for a bathing water to comply, 95 per cent of samples taken must meet these standards. In 2000, 514 of the 545 coastal waters passed the coliform standards, as did 9 of the 11 inland waters (only 9 were tested).

Published UK guidance/recommendations

ACMSF: No specific recommendations.

Pennington: There should be an education/awareness programme for farm workers, repeated and updated periodically as appropriate to ensure that they are aware of the existence, potential prevalence and nature of *E.coli* O157, the potential for the spread of infection on farms in a number of ways, including notably from faecal material and of the need for care in the use of untreated slurry or manure.

Evidence presented to VTEC taskforce: Subjects covered included: clarification of standards for composting, manure storage and treatment, proactively promote Control of Pollution Regulations, Prevention of Environmental Pollution from Agricultural Activity Code and Farm Waste Management Plans; initiate awareness raising campaign on waste storage and application for farmers and contractors, change perceptions; economic support needed for farmers; encourage waste management training for farmers and contractors; minimise volumes and leakage of contaminated water.

WHO: Regulatory authorities should be made aware that EHEC infections have been attributed to swimming/playing in (untreated) water contaminated by a variety of sources.

Key points and practical issues

- It has been shown that VTEC O157 can survive for at least 12 weeks in fresh surface water. It can enter inland and coastal waters in run-off from agricultural land. Many coastal waters are designated as bathing waters and the quality of the water is assessed regularly during the bathing season. Only a small number of inland waters have been designated as bathing waters in the UK and bathing in rivers, streams, etc. is mainly confined to the summer months. However, many more rivers and lakes are used for water sports such as canoeing, sailing and wind-surfing which are carried out throughout the year and during which surface water can be swallowed. Several outbreaks of VTEC O157 associated with bathing in recreational water have been reported. Case control studies of sporadic cases in England and the United States have also found an association with contact with recreational water. The source of the initial contamination cannot always be determined and may be either animals or an infected swimmer. In 1999 an outbreak was identified in South Devon associated with using a particular stretch of beach. The exact source of infection could not be determined although an infected child, animal or bird was suspected. As part of the outbreak investigations VTEC O157 of a 'phage type different from the outbreak strain was detected in river water from the estuary indicating the river water is certainly a potential source of contamination for beaches and coastal bathing waters. Isolating VTEC O157 from recreational water implicated in outbreaks may be difficult as the organism may be present in only very low numbers, therefore enrichment techniques together with immuno-magnetic separation should be attempted.

Further reading

Ackman, D., Marks, S., Mack, P., Caldwell, M., Root, T. and Birkhead, G. Swimming-associated haemorrhagic colitis due to Escherichia coli O157:H7 infection: evidence of prolonged contamination of a fresh water lake. *Epidemiology & Infection* 1997; 119: 1–8.

Cransberg, K., van den Kerkhof, J.H, Banffer, J.R., Stijnen, C. and Wernars, K. van *et al*. Four cases of hemolytic uremic syndrome – source contaminated swimming water? *Clinical Nephrology* 1996; 46: 45–9.

Outbreak of Escherichia coli O157:H7 and Campylobacter among attendees of the Washington County Fair – New York, 1999. *MMWR – Morbidity & Mortality Weekly Report* 1999; 48: 803–5.

Paunio, M., Pebody, R., Keskimaki, M., Kokki, M., Ruutu, P. and Oinonen, S. *et al*. Swimming-associated outbreak of Escherichia coli O157:H7. *Epidemiology & Infection* 1999; 122: 1–5.

Slutsker, L., Ries, A.A., Maloney, K., Wells, J.G., Greene, K.D. and Griffin, P.M. A nationwide case-control study of Escherichia coli O157:H7 infection in the United States. *Journal of Infectious Diseases* 1998; 177: 962–6.

VTEC O157 outbreak linked to beach holidays. *Communicable Disease Report* 330; *CDR Weekly* 9: 327.

CONTROL POINT 8: CONTAMINATION OF DRINKING WATER BY HUMAN FAECES

Aim: To prevent human sewage containing VTEC O157 from contaminating drinking water supplies

Agencies: MAFF, the Environment Agency, local authority environmental health departments, Drinking Water Inspectorate, farmers' unions and representative organisations (e.g. NFU), FSA, Water Industry.

Legislation

Environmental Protection Act 1990.
Water Resources Act 1991.
Water Industry Act 1991.
The Private Water Supplies Regulations 1991.
Natural Mineral Water Regulations 1985.
The Water Supply (Water Quality) Regulations 1989 (as amended).
The Surface Waters (Abstraction for Drinking Water) (Classification) Regulations 1996.
Environmental Protection Act 1990.

Waste Management Licensing Regulations 1994.
Sludge (Use in Agriculture) Regulations 1989.

Enforcement arrangements

The microbiological quality of surface water intended for the abstraction of drinking water, for bathing water, freshwater fish, shellfish and for water intended for human consumption is governed by European directives, such as that governing surface water (Directive 75/440/EEC), bathing water (Directive 76/160/EEC) and water intended for human consumption (Directive 80/778). The directives are implemented through legislation.

The Environment Agency monitors and controls pollution of groundwater and watercourses. Drinking water is supplied in England and Wales by private companies, in Scotland by publicly owned water authorities, and in Northern Ireland by the Northern Ireland Water Service. Section 70 of the Water Industry Act 1991 requires the water companies to supply water that is fit for human consumption.

In England and Wales, the performance of water companies in providing wholesome water complying with the Water Supply (Water Quality) Regulations is monitored by the Drinking Water Inspectorate. The DWI carries out technical audits of water companies which include assessments of sampling and progress with improvement programmes. Where companies are in breach of the regulations, the DWI issues a notification that enforcement action is being considered. If no evidence is provided that remedial action to prevent a recurrence has been taken, an undertaking to carry out works by a specified date is accepted from the company or an enforcement order is served.

The DWI also investigates incidents where water companies have supplied water that was unfit for human consumption and can bring prosecutions for this offence. Summaries of improvement programmes and enforcement action together with incidents and prosecutions are publicly available on the DWI website.

Sources of contamination for water supplies from human sewage are cesspits and septic tanks, leaking or broken sewage pipes and sewage sludge applied to agricultural land as fertiliser.

The Water Resources Act 1991 prohibits the known discharge of poisonous, noxious or polluting matter or solid waste into any 'controlled water'. 'Controlled waters' include groundwaters and all inland waters including lakes, ponds, rivers, streams, canals and field ditches. The legislation is enforced by the Environment Agency and fines and imprisonment can be imposed as well as costs for prosecution and remedial work.

Local authorities (in practice the environmental health department) are required, by the Water Industry Act 1991, to keep themselves informed about the wholesomeness of public and private water supplies in their area. In the case of private water supplies, supplied for domestic purposes or for food production to staff canteens, hostels, etc. local authorities have the responsibility to ensure that they are wholesome and can serve notices under the Water Industry Act requiring improvements. The local authorities' duties with regard to safeguarding the quality of private water supplies are specified by the Private Water Supplies Regulations 1991, and with regard to the quality of bottled mineral water in the Natural Mineral Water Regulations 1985. The regulations prescribe standards for judging the wholesomeness of private water supplies including a standard of no coliforms, faecal coliforms or faecal streptococci per 100 ml. Local authorities have a duty to take samples from all private supplies serving premises in their area. The sampling frequency is determined by the use of the supply (domestic only or hospital, canteen, etc.) and the number of persons supplied and average daily volume of water supplied.

The 1994 regulations apply to the disposal of non-agricultural waste by spreading on land. Such activities are licensed and monitored by the Environment Agency. Small amounts (no more than 250 tonnes per hectare in any 12 month period) of blood and gut contents from abattoirs and septic tank sludge, both potential sources of VTEC O157, can be applied to agricultural land without licence so long as certain conditions are met and that the Environment Agency is notified.

More than half of the sewage sludge produced in England and Wales is applied to agricultural land as organic fertiliser. Further quantities are used in forestry and land restoration. The Environment Agency monitors the provisions of the regulations which include measures to protect crops intended for human consumption. The harvesting of forage crops or the grazing of animals on agricultural land to which treated sludge has been surface applied is prohibited for at least three weeks after application. Fruit or other crops normally eaten raw must not be harvested for at least ten months after sewage sludge has been applied.

The 1989 regulations are due to be amended to include the Agricultural Development Advisory Service safe sludge matrix. As from 31 December 1999, untreated sludges cannot be applied to food crops. As of 31 December 1998, conventionally treated sewage sludge can only be applied to grazed land if it is deep injected into the soil and if there is no grazing or harvesting within 3 weeks. Conventionally treated sludge (usually by anaerobic digestion which kills 99 per cent of contaminating pathogens) can be applied to land used to grow vegetables (including potatoes, parsnips and leeks) at least 12 months before a crop is harvested. Where the crop might be eaten raw (lettuce, onions, carrots, cabbage, etc.) the harvest interval must be at least 30 months. Where enhanced treated sludges are used (capable of producing a 6 log reduction in pathogens) a 10 month harvest interval is required.

Published UK guidance/recommendations

ACMSF: No specific recommendations.

Pennington: There should be an education/awareness programme for farm workers, repeated and updated periodically as appropriate to ensure that they are aware of the existence, potential prevalence and nature of *E.coli* O157, the potential for the spread of infection on farms in a number of ways, including notably from faecal material, and of the need for care in the use of untreated slurry or manure.

Evidence presented to VTEC taskforce: More control of imported off-farm wastes, sewage sludges to be enhanced treated in accordance with ADAS matrix and tested prior to recycling to land, investigate presence of VTEC O157 in discharges to at-risk sewers.

Continuous refinement of methodology, consideration of risks to public supplies when water treatment process or chlorination failures arise or when treated water is potentially contaminated; communication of *E.coli* and enterococci detections to relevant stakeholders; upgrade/replace poor supplies related to degree of risk and improve raw water source protection; research required into survival of O157 in water and removal by water treatment process, urgent review of the private water supplies regulations (e.g. sampling frequencies and treatment standards); compulsory treatment of private water supplies for transient and tourist populations; national protocol for action when O157 detected in private or public supplies; standardisation of boil notices/advice; further research into human immunity/susceptibility to O157; education of users on risks associated with faecal contamination of supplies; financial assistance with provision of suitable water treatment systems; encourage solicitors/mortgage providers to include water supplies as part of property survey; increase availability of mains water supply; include water supplies in building regulations; find out about your water supply and responsibility for upkeep; keep private water supply safe by protection from animals, run-off, and other possible sources of contamination, if there is suspicion that private water supply is unsafe, then water should be boiled, install treatment facilities, use bottled water, or connect to mains supply; consider other issues, e.g. lead, chemicals and microorganisms when treating a private supply; when on holiday, check whether water is from a private supply, if so, boil water for drinking and food production purposes or use bottled water; application of the microbiological risk assessment protocol to all private water supplies; provision of co-ordinated education/support for owners of private water supplies to aid them in the maintenance of such supplies; continued development of the waterborne disease surveillance programme within SCIEH; continued co-operation between SCIEH and

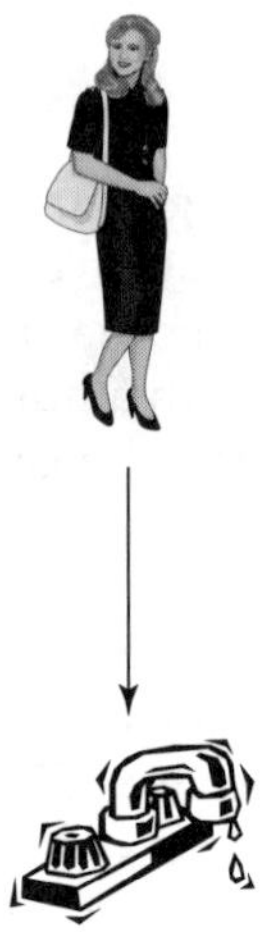

Figure 7.9 Control Point 8: Contamination of drinking water by human faeces

PHLS through strengthened links between these two organisations; private water supplies should be tested regularly for faecal coliforms, especially those in areas where cattle graze or slurry or manure are spread. Attempt to limit access of wild animals/birds to human water supplies especially untreated supplies.

Interim guidance on recreational use of animal pasture. Ensure that water from burns and streams is treated before drinking.

WHO: Regulatory authorities should be made aware that EHEC infections have been attributed to drinking untreated well-water contaminated by a variety of sources; and they should take appropriate action.

Key points and practical issues

- About 18,000 million litres of drinking water are supplied daily in the UK via 397,401 km of water mains. Drinking water is drawn from 1,584 boreholes, 666 reservoirs and 602 river abstractions. Two-thirds of our drinking water comes from surface water, the remainder from groundwater. There are around 2,250 treatment plants where water undergoes a series of processes which may include screening, coagulation, clarification, filtration, pH correction and a final disinfection with chlorine. The majority of drinking water in England and Wales is supplied by private water companies. About 1 per cent of the population of England and Wales have a private water supply, not supplied by a water company, which are generally located in rural areas and which serve only one or a limited number of properties. The source of the supply is usually a spring, well, borehole, stream, river, lake or pond. In general, springs, boreholes and wells are less likely to become contaminated as they draw water from deep underground although contamination can still occur at points of emergence or collection.

- Sewage sludge arises from the treatment of human sewage. Untreated sludge is obtained from the settlement or biological stages of treatment. Further processing can be undertaken to improve its stability. It is applied to land as semi-liquid (similar to animal slurry) or as de-watered sludge cake.
- Sludge from septic tanks and cesspools are also commonly spread on agricultural land. Septic tank waste is partially digested sewage and can be spread on agricultural land with the same controls as applied to sewage sludge. Cesspool waste is raw and undigested. Spreading cesspool waste must be licensed by the Environment Agency under the 1994 regulations.
- It has been shown that VTEC O157 can survive for at least 12 weeks in water. Microbial contamination of such supplies is particularly likely during periods of high rainfall (more run-off) and warm weather (low supplies). Various methods of disinfecting private water supplies including ultraviolet disinfection units are available. Research has shown that it is necessary to apply a UV fluence of at least 400 J/m2 to destroy VTEC O157. Electrolysed oxidising has also been suggested as an alternative disinfection method.
- Untreated (private) supplies are particularly vulnerable and found mainly in agricultural areas. However, public chlorinated supplies can also be the source of outbreaks where the integrity of the supply network is breached, for example, during repairs or breakages. Several outbreaks, both in the UK and abroad, have been attributed to contamination of water supplies by human sewage. The route of contamination was seepage from sewage pipes and overflows into water supply pipes when repairs to the water pipes caused a loss of positive pressure. In a further outbreak, a water supply became contaminated by stream water. The sewage outflow for the village was upstream of the point where the stream water had been abstracted. Waterborne outbreaks can be difficult to investigate. Contamination of the supply can be intermittent and bacteria levels in water can be very low making it almost impossible to isolate the causative organism. In one outbreak, bacteriological tests for *E.coli* were negative whereas VTEC O157 was isolated using Sorbitol MacConkey Agar. Specific methods for isolating VTEC O157 from water can be used. Careful epidemiological investigation of cases by gathering information on sources of water supplies

and water consumption histories is useful. Information on sewage and water pipe breaks and repairs is also relevant. Case control studies of sporadic cases have also identified consumption of unchlorinated well water as a risk factor.

Further reading

Kurokawa, K., Tani, K., Ogawa, M. and Nasu, M. Abundance and distribution of bacteria carrying sltII gene in natural river water. *Letters in Applied Microbiology* 1999; 28: 405–10.

Outbreak of Escherichia coli O157:H7 and Campylobacter among attendees of the Washington County Fair – New York, 1999. *MMWR – Morbidity & Mortality Weekly Report* 1999; 48: 803–5.

Slutsker, L., Ries, A.A., Maloney, K., Wells, J.G., Greene, K.D. and Griffin, P.M. A nationwide case-control study of Escherichia coli O157:H7 infection in the United States. *Journal of Infectious Diseases* 1998; 177: 962–6.

Sommer, R., Lhotsky, M., Haider, T. and Cabaj, A. UV inactivation, liquid-holding recovery, and photoreactivation of Escherichia coli O157 and other pathogenic Escherichia coli strains in water. *Journal of Food Protection* 2000; 63: 1015–20.

Tanaka, Y., Yamaguchi, N. and Nasu, M. Viability of Escherichia coli O157:H7 in natural river water determined by the use of flow cytometry. *Journal of Applied Microbiology* 2000; 88: 228–36.

Venkitanarayanan, K.S., Ezeike, G.O., Hung, Y.C. and Doyle, M.P. Efficacy of electrolyzed oxidizing water for inactivating Escherichia coli O157:H7, Salmonella enteritidis, and Listeria monocytogenes. *Applied & Environmental Microbiology* 1999; 65: 4276–9.

Wang, G. and Doyle, M.P. Survival of enterohemorrhagic Escherichia coli O157:H7 in water. *Journal of Food Protection* 1998; 61: 662–7.

CONTROL POINT 9: CONTAMINATION OF RECREATIONAL WATER BY HUMAN FAECES

Aim: To prevent human sewage containing VTEC O157 from contaminating recreational water

Agencies: MAFF, the Environment Agency, local authority environmental health departments, farmers' unions and representative organisations (e.g. NFU).

Legislation

Environmental Protection Act 1990.

Water Resources Act 1991.
Water Industry Act 1991.
Health and Safety at Work etc. Act 1974.
Health and Safety at Work Regulations 1992.

Enforcement arrangements

The Environment Agency monitors and controls pollution of groundwater and watercourses. Sewage discharges to coastal and surface waters must be approved by the Environment Agency and required standards are set. Water companies or other organisations found to be in breach of these standards can be fined under section 85 of the Water Resources Act 1991 and required to pay the clean-up costs. Untreated sewage is most likely to enter watercourses during periods of heavy rain when emergency or storm overflows are used to prevent sewage works becoming overloaded.

Sources of contamination for water supplies from human sewage are cesspits and septic tanks, leaking or broken sewage pipes and sewage sludge applied to agricultural land as fertiliser.

The Water Resources Act 1991 prohibits the known discharge of poisonous, noxious or polluting matter or solid waste into any 'controlled water'. 'Controlled waters' include groundwaters and all inland waters including lakes, ponds, rivers, streams, canals and field ditches. The legislation is enforced by the Environment Agency and fines and imprisonment can be imposed on farmers, employees or contractors as well as costs for prosecution and remedial work.

The microbiological quality of surface water intended for bathing is governed by European Directive 76/160/EEC. The directives are implemented through UK legislation. Directive 76/160/EEC requires member states to identify relevant bathing waters. There are currently 11 inland waters and 545 coastal waters identified as bathing waters.

The annual assessment of compliance with the Bathing Waters Directive is made by the Department of the Environment, Transport and the Regions (DETR). In practice monitoring is completed by the Environment Agency by taking samples of water. The bathing season in England and Wales runs from 15 May to 30 September and sampling commences two weeks before the start of the season. Twenty samples are taken at regular intervals throughout the season at each site. Samples are taken at points where the average density of bathers is at its highest. There must be no more than 10,000 total coliforms per 100 ml and no more than 2,000 faecal coliforms per 100 ml. In order for a bathing water to comply, 95 per cent of samples taken must meet these standards. In 2000, 514 of the 545 coastal waters passed the coliform standards, as did 9 of the 11 inland waters (only 9 were tested).

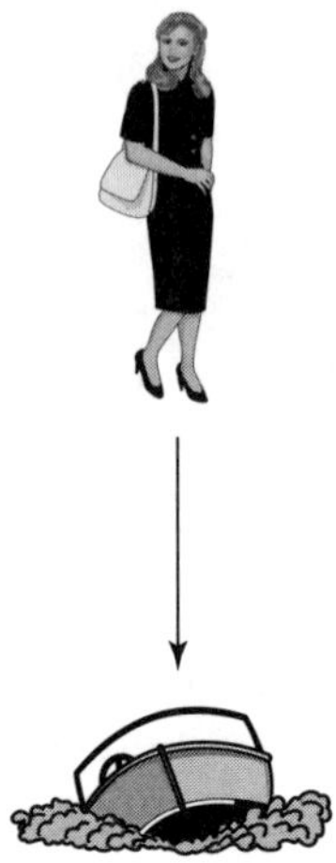

Figure 7.10 Control Point 9: Contamination of recreational water by human faeces

The 1994 regulations apply to the disposal of non-agricultural waste by spreading on land. Such activities are licensed and monitored by the Environment Agency. Small amounts (no more than 250 tonnes per hectare in any 12 month period) of blood and gut contents from abattoirs and septic tank sludge, both potential sources of VTEC O157, can be applied to agricultural land without licence so long as certain conditions are met and that the Environment Agency is notified.

More than half of the sewage sludge produced in England and Wales is applied to agricultural land as organic fertiliser. Further quantities are used in forestry and land restoration. The Environment Agency monitors the provisions of the regulations which include measures to protect crops intended for human consumption. The 1989 regulations are due to be amended to include the Agricultural Development Advisory Service safe sludge matrix. As of 31 December 1998, conventionally treated sewage sludge can only be applied to grazed land if it is deep injected into the soil and if there is no grazing or harvesting within 3 weeks. Conventionally treated sludge (usually by anaerobic digestion which kills 99 per cent of contaminating pathogens) can be applied to land used to grow vegetables (including potatoes, parsnips and leeks) at least 12 months before a crop is harvested. The harvesting of forage crops or the grazing of animals on agricultural land to which treated sludge has been surface applied is prohibited for at least 3 weeks after application. Fruit or other crops normally eaten raw must not be harvested for at least 10 months after sewage sludge has been applied.

Local authorities also play a role in maintaining the safety of recreational water. Section 259 of the Public Health Act 1936 makes any pond or watercourse which is so foul, or in such a state to be prejudicial to health, a statutory nuisance. Notices can be served to secure improvements.

The monitoring and control of the bacteriological quality of swimming pool water is also the responsibility of environmental health departments, with powers to require improvements provided either by model bylaws under the Public Health Act 1936 (if adopted) or the Health and Safety at Work etc. Act 1974 (Section 3 referring to 'other persons affected by the work activity) and the Management of Health and Safety at Work Regulations 1992, which require an employer to make an assessment of risks of hazards to employees and other persons.

Published UK guidance/recommendations

ACMSF: No specific recommendations.

Pennington: There should be an education/awareness programme for farm workers, repeated and updated periodically as appropriate to ensure that they are aware of the existence, potential prevalence and nature of *E.coli* O157, the potential for the spread of infection on farms in a number of ways, including notably from faecal material and of the need for care in the use of untreated slurry or manure.

Evidence presented to VTEC taskforce: More control of imported off-farm wastes, sewage sludges to be enhanced treated in accordance with ADAS matrix and tested prior to recycling to land, investigate presence of VTEC O157 in discharges to at-risk sewers.

WHO: Regulatory authorities should be made aware that EHEC infections have been attributed to swimming/playing in (untreated) water contaminated by a variety of sources.

Key points and practical issues

- It has been shown that VTEC O157 can survive for at least 12 weeks in fresh surface water. It can enter inland and coastal waters in run-off from agricultural land fertilised by sewage sludge, from cesspits and septic tanks and from leaking sewers. Many coastal waters are designated as bathing waters and the quality of the water is assessed regularly during the bathing season. Only a small number of inland waters have been designated as bathing waters in the UK and bathing in rivers, streams, etc. is mainly confined to the summer months. However, many more rivers and lakes are used for water sports such as canoeing, sailing and wind-surfing which are carried out throughout the year and during which surface water can be swallowed. Several outbreaks of VTEC O157 associated with bathing in recreational water have been reported. Case control studies of sporadic cases in England and the United States have also found an association with contact with recreational water. The source of the initial contamination cannot always be determined and may be either animals or an infected swimmer. In 1999 an outbreak was identified in South Devon associated with using a particular stretch of beach. The exact source of infection could not be determined although an infected child, animal or bird was suspected. Isolating VTEC O157 from recreational water implicated in outbreaks may be difficult as the organism may be present in only very low numbers therefore enrichment techniques together with immunomagnetic separation should be attempted.

- The main source of contamination of swimming pools in the UK is an infected swimmer. Swimming pools frequented by children are obviously more likely to be contaminated by faecal material. Outbreaks associated with private small paddling pools used by several children in a back garden have been reported. Commercially run swimming pools require adequate filtration and treatment (chlorination) to ensure that VTEC O157 cannot survive in the water.
- Outdoor pools can also potentially be contaminated by wildlife and by soil walked into the pool on the feet of bathers.

Further reading

Ackman, D., Marks, S., Mack, P., Caldwell, M., Root, T. and Birkhead, G. Swimming-associated haemorrhagic colitis due to Escherichia coli O157:H7 infection: evidence of prolonged contamination of a fresh water lake. *Epidemiology & Infection* 1997; 119: 1–8.

Cransberg, K., van den Kerkhof, J.H., Banffer, J.R., Stijnen, C. and Wernars, K. van *et al.* Four cases of hemolytic uremic syndrome – source contaminated swimming water? *Clinical Nephrology* 1996; 46: 45–9.

Outbreak of Escherichia coli O157:H7 and Campylobacter among attendees of the Washington County Fair – New York, 1999. *MMWR – Morbidity & Mortality Weekly Report* 1999; 48: 803–5.

Paunio, M., Pebody, R., Keskimaki, M., Kokki, M., Ruutu, P. and Oinonen, S. *et al.* Swimming-associated outbreak of Escherichia coli O157:H7. *Epidemiology & Infection* 1999; 122: 1–5.

Slutsker, L., Ries, A.A., Maloney, K., Wells, J.G., Greene, K.D. and Griffin, P.M. A nationwide case-control study of Escherichia coli O157:H7 infection in the United States. *Journal of Infectious Diseases* 1998; 177: 962–6.

VTEC O157 outbreak linked to beach holidays. *Communicable Disease Report* 330; *CDR Weekly* 1999; 9: 327.

CONTROL POINT 10: ZOONOTIC SPREAD

Aim: To prevent direct spread of VTEC O157 to humans via animal manure

Agencies: MAFF, SVS, HSE, local authorities, farmers' unions and representative organisations (e.g. NFU).

Legislation

Health and Safety at Work Act 1974.
Health and Safety (Enforcing Authority) Regulations, 1998.
Control of Substances Hazardous to Health Regulations 1999.

Enforcement arrangements

The control of infection on farms open to the public is governed by the Health and Safety at Work etc. Act 1974, the Management of Health and Safety Regulations 1992, and the Control of Substances Hazardous to Health Regulations 1999. The Health and Safety Executive enforce the legislation on premises where the main purpose is agriculture. Where the main purpose is entertainment (as in the case of many open farms) enforcement of this legislation is the responsibility of the local authority environmental health departments. Enforcement is via improvement or prohibition notices and fines. The COSHH Regulations 1999 require employers and self-employed people to assess the risks to health from work activities which involve a hazardous substance (e.g. a microorganism); prevent or, when this is not reasonably practicable, adequately control exposure to the hazardous substances; introduce and maintain control measures; inform, instruct and train employees about the risks and precautions to be taken; regularly review risk assessments and the effectiveness of control measures. Specific recommendations for VTEC O157 are thorough hand and arm washing before eating, drinking or smoking and leaving workwear at the workplace for cleaning.

Written guidance is available for teachers accompanying children to open farms. Ratios for pupil supervision, appropriate clothing, conduct around animals (no food, sweets, thumb sucking) and hand washing are included.

Written guidance for farmers operating open farms includes recommendations for layout, animal contact, siting of eating areas, washing facilities, information and signs, training and supervision, livestock management procedures and manure and compost heaps.

Further advice is available for farmers and their employees on farms not open to the public.

Published UK guidance/recommendations

ACMSF: Recommendation for research into epidemiology of VTEC O157 in agricultural livestock and husbandry and other factors contributing to herd infection and control.

Pennington: Education/awareness programme for farm workers, repeated and updated periodically to ensure that they are aware:

(a) of the existence, potential prevalence and nature of *E.coli* O157
(b) of the potential for the spread of infection on farms in a number of ways, including notably from faecal material, and of the consequent need for scrupulous personal hygiene.

Evidence presented to VTEC taskforce: Open farms: One-stop-shop required for compliance and advice, with practical guidance and checklists for open farms; publicise advice through farm park/attractions, associations and farmers' unions, FFES, local and farming media, etc.; include country parks (including local authority) and all visitor facilities with animals; advise how to reduce the amount of infection; encourage all animal visitor facilities to make hygiene part of the educational experience; provide certificates of compliance/safe practice for open farms; express the risk at farm parks fairly and comparatively. Health risks should be compared with beaches, public parks, leisure centres, swimming pools; provide guidance for local authorities/schools, etc. so they don't need to over-react; give local authorities protection against liability if they observe FSA guidance.

Interim guidance on recreational use of animal pasture: Ideally to completely avoid risk of infection by *E.coli* O157 from this source, fields used for grazing or stockholding of animals should not be used for camping, picnicking and play areas, especially where these involve children.

However, as with everyday life, these risks can be greatly reduced by adopting the following sensible precautions: keep farm animals off the fields for the preceding 3 weeks prior to use; keep farm animals off fields during use; remove any visible dropping, ideally at the beginning of the 3 week period; mow the grass, keep it short and remove the clippings before the fields are used for recreation; always wash hands before eating, drinking and smoking, i.e. use soap, clean towels and preferably, hot running water; ensure adequate supervision of children, particularly those under 5 years of age.

WHO: Both adults and children having animal contact on farms and at other sites should be made aware of the extremely low infective dose of *E.coli* O157 and possibly other EHEC, and the potential for its transmission

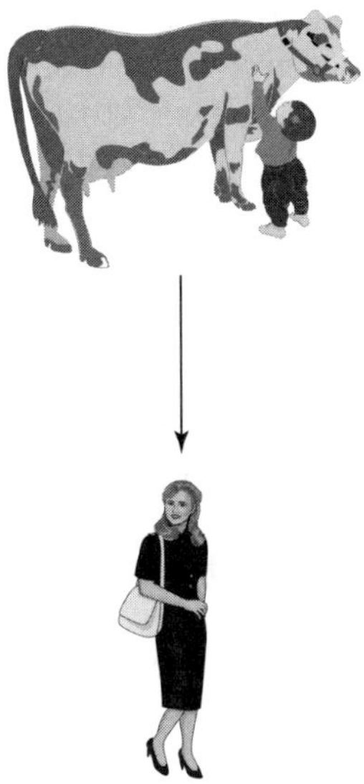

Figure 7.11 Control Point 10: Zoonotic spread

from such contact. Measures to implement and enhance strict hygiene practices in these settings should be encouraged. Visitors to farms, who may be unaware of the highly infectious nature of EHEC in these settings should be instructed to maintain good handwashing and other hygienic practices. This is of particular importance for children, as several outbreaks have occurred among school children after visiting farms.

Key points and practical issues

- There are basically three situations in which humans are exposed to animal manure. The first and most obvious is direct contact with animals and their manure. Farmers and other agricultural workers are exposed to this route daily. Members of the public are exposed when they visit either private or open farms and stroke or touch the animals. Second, animal manure readily contaminates the surrounding environment as it is frequently very liquid in consistency. Barns, shed gates and fences can be contaminated as can pasture used for grazing. This contamination can be difficult to see and it is possible to be infected with VTEC O157 by merely being in contact with these contaminated articles without actually touching an animal or knowingly touching manure. Coats and boots worn when handling animals can carry infection into the farm house and infection can be carried to locations further afield via vehicles, etc. Agricultural vehicles and buildings can be a source of infection for mechanics and electricians, builders, and others visiting the farm for work purposes. Case control studies of sporadic cases have identified this as a risk factor. Recreational use of land used by animals for grazing is a further route. Individuals are exposed to this route when walking or camping and playing sports on agricultural land or when attending shows or festivals held on pasture previously grazed by animals. It is likely that wet conditions creating a lot of mud increase the chances of infection being transmitted by this route. Finally, animal manure is removed from farms and used as fertiliser in horticulture where it may be applied directly to domestic gardens or stored in compost heaps, etc. This presents a further route of infection without the need for direct animal contact.

- Several species of animal found at open farms, including cattle, goats, sheep, pigs, horses and deer, can excrete VTEC O157 and act as a source of infection to humans.

Further reading

Crampin, M., Willshaw, G., Hancock, R., Djuretic, T., Elstob, C. and Rouse, A. *et al.* Outbreak of Escherichia coli O157 infection associated with a music festival. *European Journal of Clinical Microbiology & Infectious Diseases* 1999; 18: 286–8.

Parry, S.M., Salmon, R.L., Willshaw, G.A. and Cheasty, T. (1998) Risk factors for and prevention of sporadic infections with vero cytotoxin (shiga toxin) producing Escherichia coli O157. *The Lancet* 351(9108): 1019–22.

Pritchard, G.C., Willshaw, G.A., Bailey, J.R., Carson, T. and Cheasty, T. Verocytotoxin-producing Escherichia coli O157 on a farm open to the public: outbreak investigation and longitudinal bacteriological study. *Veterinary Record* 2000; 147: 259–64.

Randall, L.P., Wray, C. and Davies, R.H. Survival of verocytotoxin-producing Escherichia coli O157 under simulated farm conditions. *Veterinary Record* 1999; 145: 500–1.

Two outbreaks of Vero cytotoxin producing Escherichia coli O157 infection associated with farms. *Communicable Disease Report* 266; *CDR Weekly* 7: 263.

VTEC O157 outbreak associated with a farm visitor centre in North Wales. *Communicable Disease Report* 230; *CDR Weekly* 9: 227.

VTEC O157 outbreak associated with a farm visitor centre in North Wales – update. *Communicable Disease Report* 248; *CDR Weekly* 9: 245.

CONTROL POINT 11: CONSUMPTION OF RAW AND UNDERCOOKED MEAT AND OTHER FOOD PRODUCTS

Aim: To prevent survival of VTEC O157 in meat and other food products to the point of consumption

Agencies: FSA, local authorities, retail food and catering industry organisations.

Legislation

Food Safety Act 1990.
Food Safety (General Food Hygiene) Regulations 1995.

Enforcement arrangements

The Food Safety Act 1990 makes it an offence to sell food for human consumption which fails to comply with food safety requirements. A food fails to comply with food safety requirements if it has been rendered injurious to health, is unfit for human consumption or is so contaminated that it would not be reasonable to expect it to be used for human consumption. The legislation is enforced by authorised officers (Environmental Health Officers and other authorised officers) working for a food authority (local authorities). Officers have the power to inspect any food intended for human consumption and to order the detention or seizure or food by giving notice to the person in charge of the food. Where food is seized, it is brought before a justice of the peace, who has the power to condemn the food and order that it be destroyed so as to prevent it from being used for human consumption.

The Food Safety (General Food Hygiene) Regulations 1995, are made under the Food Safety Act and implement Council Directive 93/43/EEC on the hygiene of foodstuffs. They apply to all stages of food production except primary production. The regulations require proprietors of food businesses to identify any step in the activities of the food business which is critical to ensuring food safety and ensure that adequate safety procedures are identified, implemented, maintained and reviewed on the basis of analysis of hazards; identification of points where hazards may occur; deciding on critical points; identification and implementation of effective control and monitoring of those points and review.

A person found guilty of contravening the above can incur a fine or imprisonment up to a term of two years. If, following a conviction, the court is satisfied that a 'health risk condition' is satisfied, the court can prohibit the use of a process or treatment, equipment or premises. The 'health risk condition' is deemed to be fulfilled if any process, treatment, premises or equipment involves risk of injury to health.

Failure to comply with the regulations can be addressed via an improvement notice served on the proprietor specifying the measures which must be taken to secure compliance and a period within which they must be carried out.

Authorised officers can also serve emergency prohibition notices where the 'health risk condition' is fulfilled (see above) but where the risk of injury is imminent.

Compliance with the food hygiene regulations is assessed during routine food hygiene inspections. The frequency of these inspections is prescribed by a Code of Practice (Code of Practice No. 9 Food Hygiene Inspections). Individual premises are assigned an inspection rating based on a number of points. The premises with the highest number of points (category A) are inspected most frequently (at least every 6 months). Points are awarded for type and method of processing, consumers at risk, level of compliance and confidence in management. An additional score of 20 should be included where there is a significant risk of food being contaminated with *E.coli* O157 or other VTEC. Significant risk means a higher probability that an incident may occur and the following matters should be taken into account in assessing the risk:

(a) the potential for contamination/cross-contamination by the specified organism
(b) survival and growth of the specified organism
(c) the existence of hazard analysis systems and confidence in their implementation including documentation and records of monitoring of controls
(d) the extent and relevance of training undertaken by managers, supervisors and food handlers
(e) whether intervention by food authorities is necessary to reduce the probability of an incident occurring.

Between 1997 and 1998 additional funding of £19 million was made available for implementing the key recommendations of the Pennington Report, i.e. increasing the level of food safety enforcement in high risk premises and accelerating the implementation of food safety control systems based on HACCP.

Published UK guidance/recommendations

ACMSF: We recommend that all relevant sectors of the food industry adopt a HACCP-based approach to prevent survival of or contamination by VTEC.

We recommend that persons preparing instructions for the cooking of beefburgers or actually cooking beefburgers must pay particular attention to the formulation of the burger; its thickness; the methods of defrosting and cooking used; and should monitor the cooking process, taking remedial action where necessary.

We recommend that the government and industry fund research and surveillance into the relationship between the formulation and colour of cooked minced meat products, the colour of juices, and the temperature and the survival of VTEC.

Revised recommendations on cooking burgers (March 1997)

General

We continue to recommend that minced meat and minced meat products, including burgers, should be cooked to a minimum temperature of 70°C for 2 minutes or equivalent.

For consumers

Consumers should follow the manufacturer's instructions. In any event, it is particularly important that consumers ensure that burgers, and similar minced meat products, are thoroughly cooked so that they are piping hot throughout. Eating undercooked burgers that are rare in the middle may be dangerous.

It should be noted that when barbecues are used, the cooking process is variable and difficult to control. This makes it doubly important to ensure that burgers are cooked thoroughly so that they are piping hot throughout.

For manufacturers and retailers

Minced meat and minced meat products, including burgers, should be cooked to a minimum temperature of 70°C for 2 minutes or equivalent. Vendors of raw burgers should ensure that all burgers and similar minced meat products are supplied with cooking instructions that are adequate to comply with this recommendation. Cooking instructions should also take into account factors such as whether the burger is frozen or chilled, the thickness and formulation of the burger, and the prescribed method of cooking.

We therefore recommend that the advice to cook burgers until the juices run clear and there are no pink bits inside may be used where appropriate (e.g. when a burger contains only beef and no added salt), but that it should always be accompanied by other cooking instructions which achieve a minimum temperature of 70°C for 2 minutes or equivalent.

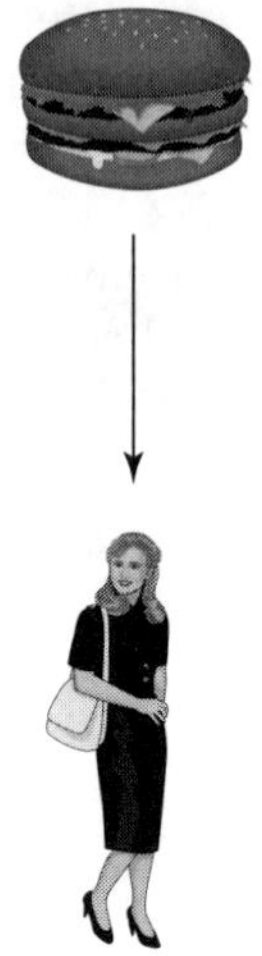

Figure 7.12 Control Point 11: Consumption of raw and undercooked foods

For caterers and catering supplies

WHOLESALE SUPPLIES TO CATERERS

We recommend that cartons of burgers (and other similar minced meat products) supplied by wholesalers for caterers should be labelled with a clear instruction that the product must always be cooked thoroughly so that it is piping hot right through to the centre. Minced meat and minced meat products, including burgers, should be cooked to a minimum temperature of 70°C for 2 minutes or equivalent.

CATERERS

Vendors of cooked burgers and other similar minced meat products, for example caterers, have a specific legal obligation to identify and control any process steps that are critical to food safety (Food Safety (General Food Hygiene) Regulations 1995, regulation 4(3)). The thorough cooking of minced meat products, including burgers, to a temperature of 70°C for 2 minutes or equivalent, will be such a critical control. Caterers must ensure that their procedures achieve this and they should take into account the type of cooking equipment, its operating temperature, the temperature of the meat product at the start of cooking, its thickness and any other relevant factors.

Caterers should consider the potential for undercooked burgers to cause disease and should not provide customers with undercooked burgers, or, if specifically requested to do so, should remind the customer of the potential hazard.

Training

VTEC infections could be significantly reduced if there were a better understanding of the need to avoid cross-contamination and to cook food properly.

We recommend commercial food handlers to focus on methods for the safe and hygienic handling of food. In catering establishments, ensure that the staff know precisely what to do – e.g. the routine for safe cooking and why it must be followed.

NB

New research indicates that heat resistance studies with *E.coli* O157 support the current advice to cook minced beef products thoroughly so that the temperature at the centre reaches 70°C for 2 minutes.

Colour alone is unlikely to be a reliable indicator that the burger has reached 70°C for 2 minutes (or an equivalent treatment).

The rate of myoglobin denaturation during heating, and hence loss of redness, will be influenced by the type of meat (lamb appears to be faster than beef), cut of beef used (sirloin faster than chuck), and the salt level present (beef faster with higher salt).

Temperature profiles of commercial burgers during cooking suggest that many appear cooked well before a minimum temperature of 70°C is reached. However, most of the burgers reached 70°C if the manufacturer's cooking instructions were followed.

Pennington: HACCP (i.e. the approach and all seven principles) should be adopted by all food businesses to ensure food safety. While this is being negotiated into European Union and domestic legislation, implementation and enforcement of the HACCP principles contained in the existing legislation should be accelerated.

The government should seek to have all the HACCP elements negotiated within the Horizontal Directive.

Evidence presented to VTEC taskforce: Issues included marketing of infection free meat, new novel food processing methods to reduce bacterial load, risk reduction throughout the food chain, advice/discussion on risk assessment in controllable (e.g food chain) circumstances, improved education/communication, use of media and pooling of expertise and experience.

WHO: For food retail and food services: Food handlers, including home caterers, should be trained in the principles of food hygiene and the application of HACCP, with particular attention to the special precautions of preparing and serving food to vulnerable groups, in accordance with WHO recommendations Hygiene in Food-Service and Mass Catering establishments (WHO/FNU/FOS/94.5); *Safe Food Handling – A Guide for Managers of Food Service Establishments*, by M. Jacob (WHO 1989).

Consumer education

Existing education programmes should be reviewed to ensure that they adequately address the low infectious dose of EHEC, the understanding of severe, long-term effects of infection, and the need for thorough cooking.

Food hygiene training should be carried out wherever possible within the primary and secondary school curriculum through the use of well-designed educational material, such as those provided by *WHO Food, Environment and Health – A Guide for Primary School Teachers*, by T. Williams, A. Moon and M. Williams (WHO 1990).

An education strategy on food safety for the consumer, conducted by a variety of organisations, including government, the food industry, consumer associations, trade associations, and consumer groups, should provide consistent, clearly understood messages, soundly based on science. Messages could be tested by focus groups. Conflicting messages must be avoided.

Educational programmes should use all means of disseminating information, including the media (newspaper, TV, radio) coverage, leaflets, lectures/lessons, training courses, labelling, in-store posters/leaflets.

The concept of risk communication involving all stakeholders (interested parties), including consumers, should be adopted to reduce EHEC infections.

For raw, ready to drink fruit or vegetable juices where safety of the product cannot be assured, a processing step which can remove or destroy EHEC, such as pasteurisation, should be applied.

Key points and practical issues

- The first recorded outbreaks of VTEC O157, in the USA in 1982, which led to its recognition as a pathogen were associated with undercooked minced beef. Since then, many outbreaks have been reported where undercooked minced beef was the vehicle of infection. Investigations of these outbreaks has frequently led to the isolation of the outbreak strain of VTEC O157 from implicated batches of burgers and the recall of large quantities of burgers. Several case control studies of sporadic cases have confirmed this route of transmission. One study clearly identified that a critical control point in the prevention of VTEC O157 infection was an appropriate method to ensure thorough cooking of beefburgers. In the UK, burgers from commercial food premises are more likely to be implicated than burgers cooked at home. This may be because commercial pressures in the catering industry result in inadequate cooking times. Faulty or badly designed equipment and poor staff training may also contribute. Some national chains use double sided grills which cook burgers from two directions and which cannot be opened until a predetermined cooking period has elapsed. A case control study in the UK showed that burgers cooked other than by this method were a risk factor for sporadic VTEC O157.

- VTEC O157 has been isolated from other raw meats including lamb and deer and all undercooked minced meat products should be regarded as a risk.
- Some meat products, including jerky and salami, are produced by a process of fermentation or drying rather than cooking. Several outbreaks associated with fermented salami and deer meat jerky have been reported particularly in the USA and Canada. Research suggests that the VTEC O157 is able to survive the acidic conditions of fermentation.

Further reading

Ansay, S.E., Darling, K.A. and Kaspar, C.W. Survival of Escherichia coli O157:H7 in ground-beef patties during storage at 2, –2, 15 and then –2 degrees C, and –20 degrees C. *Journal of Food Protection* 1999; 62: 1243–7.

Buchanan, R.L., Edelson, S.G., Snipes, K. and Boyd, G. Inactivation of Escherichia coli O157:H7 in apple juice by irradiation. *Applied & Environmental Microbiology* 1998; 64: 4533–5.

Cieslak, P.R., Noble, S.J., Maxson, D.J., Empey, L.C., Ravenholt, O. and Legarza, G. *et al.* Hamburger-associated Escherichia coli O157:H7 infection in Las Vegas: a hidden epidemic. *American Journal of Public Health* 1997; 87: 176–80.

Cody, S.H., Glynn, M.K., Farrar, J.A., Cairns, K.L., Griffin, P.M. and Kobayashi, J. *et al.* An outbreak of Escherichia coli O157:H7 infection from unpasteurized commercial apple juice. *Annals of Internal Medicine* 1999; 130: 202–9.

Dingman, D.W. Growth of Escherichia coli O157:H7 in bruised apple (Malus domestica) tissue as influenced by cultivar, date of harvest, and source. *Applied & Environmental Microbiology* 2000; 66: 1077–83.

Dingman, D.W. Prevalence of Escherichia coli in apple cider manufactured in Connecticut. *Journal of Food Protection* 1999; 62: 567–73.

Dock, L.L., Floros, J.D. and Linton, R.H. Heat inactivation of Escherichia coli O157:H7 in apple cider containing malic acid, sodium benzoate, and potassium sorbate. *Journal of Food Protection* 2000; 63: 1026–31.

Duffy, G., Riordan, D.C., Sheridan, J.J., Call, J.E., Whiting, R.C. and Blair, I.S. *et al.* Effect of pH on survival, thermotolerance, and verotoxin production of Escherichia coli O157:H7 during simulated fermentation and storage. *Journal of Food Protection* 2000; 63: 12–18.

Evrendilek, G.A., Zhang, Q.H. and Richter, E.R. Inactivation of Escherichia coli O157:H7 and Escherichia coli 8739 in apple juice by pulsed electric fields. *Journal of Food Protection* 1999; 62: 793–6.

Faith, N.G., Parniere, N., Larson, T., Lorang, T.D., Kaspar, C.W. and Luchansky, J.B. Viability of Escherichia coli O157:H7 in salami following conditioning of batter, fermentation and drying of sticks, and storage of slices. *Journal of Food Protection* 1998; 61: 377–82.

Faith, N.G., Le Coutour, N.S., Alvarenga, M.B., Calicioglu, M., Buege, D.R. and Luchansky, J.B. *et al.* Viability of Escherichia coli O157:H7 in ground and formed beef jerky prepared at levels of 5 and 20% fat and dried at 52, 57, 63 or 68 degrees C in a home-style dehydrator. *International Journal of Food Microbiology* 1998; 41: 213–21.

Faith, N.G., Wierzba, R.K., Ihnot, A.M., Roering, A.M., Lorang, T.D. and Kaspar, C.W. *et al.* Survival of Escherichia coli O157:H7 in full- and reduced-fat pepperoni after manufacture of sticks, storage of slices at 4 degrees C or 21 degrees C under air and vacuum, and baking of slices on frozen pizza at 135, 191 and 246 degrees C. *Journal of Food Protection* 1998; 61: 383–9.

Feng, P. Escherichia coli serotype O157:H7: novel vehicles of infection and emergence of phenotypic variants. [Review] [44 refs]. *Emerging Infectious Diseases* 1995; 1: 47–52.

From the Centers for Disease Control and Prevention. Outbreak of Escherichia coli O157:H7 infections associated with drinking unpasteurized commercial apple juice – British Columbia, California, Colorado, and Washington, October 1996. *JAMA* 1996; 276: 1865.

From the Centers for Disease Control and Prevention. Outbreaks of Escherichia coli O157:H7 infection and cryptosporidiosis associated with drinking unpasteurized apple cider – Connecticut and New York, October 1996. *JAMA* 1997; 277: 781–2.

Hilborn, E.D., Mshar, P.A., Fiorentino, T.R., Dembek, Z.F., Barrett, T.J. and Howard, R.T. *et al.* An outbreak of Escherichia coli O157:H7 infections and haemolytic uraemic syndrome associated with consumption of unpasteurized apple cider. *Epidemiology & Infection* 2000; 124: 31–6.

Janisiewicz, W.J., Conway, W.S. and Leverentz, B. Biological control of postharvest decays of apple can prevent growth of Escherichia coli O157:H7 in apple wounds. *Journal of Food Protection* 1999; 62: 1372–5.

Juneja, V.K., Snyder, O.P. Jr. and Marmer, B.S. Thermal destruction of Escherichia coli O157:H7 in beef and chicken: determination of D- and z-values. *International Journal of Food Microbiology* 1997; 35: 231–7.

Keene, W.E., Sazie, E., Kok, J., Rice, D.H., Hancock, D.D. and Balan, V.K. *et al.* An outbreak of Escherichia coli O157:H7 infections traced to jerky made from deer meat. *JAMA* 1997; 277: 1229–31.

Outbreak of Escherichia coli O157:H7 infections associated with drinking unpasteurized commercial apple juice – British Columbia, California, Colorado, and Washington, October 1996. *MMWR – Morbidity & Mortality Weekly Report* 1996; 45: 975.

Outbreaks of Escherichia coli O157:H7 infection and cryptosporidiosis associated with drinking unpasteurized apple cider – Connecticut and New York, October 1996. *Canada Communicable Disease Report* 1997; 23: 37–40.

Outbreaks of Escherichia coli O157:H7 infection associated with drinking unpasteurized apple cider – October 1996. Update on emerging infections from the Centers for Disease Control and Prevention. *Annals of Emergency Medicine* 1997; 29: 645–6.

Outbreaks of Escherichia coli O157:H7 infection and cryptosporidiosis associated with drinking unpasteurized apple cider – Connecticut and New York, October 1996. *MMWR – Morbidity & Mortality Weekly Report* 1997; 46: 4–8.

Parry, S.M., Salmon, R.L., Willshaw, G.A. and Cheasty, T. (1998) Risk factors for and prevention of sporadic infections with vero cytotoxin (shiga toxin) producing Escherichia coli O157. *The Lancet* 351 (9108): 1019–22.

Riordan, D.C., Duffy, G., Sheridan, J., Eblen, B.S., Whiting, R.C. and Blair, I.S. *et al.* Survival of Escherichia coli O157:H7 during the manufacture of pepperoni. *Journal of Food Protection* 1998; 61: 146–51.

Roering, A.M., Luchansky, J.B., Ihnot, A.M., Ansay, S.E., Kaspar, C.W. and Ingham, S.C. Comparative survival of Salmonella typhimurium DT 104, Listeria monocytogenes, and Escherichia coli O157:H7 in preservative-free apple cider and simulated gastric fluid. *International Journal of Food Microbiology* 1999; 46: 263–9.

Semanchek, J.J. and Golden, D.A. Influence of growth temperature on inactivation and injury of Escherichia coli O157:H7 by heat, acid, and freezing. *Journal of Food Protection* 1998; 61: 395–401.

Splittstoesser, D.F., McLellan, M.R. and Churey, J.J. Heat resistance of Escherichia coli O157:H7 in apple juice. *Journal of Food Protection* 1996; 59: 226–9.

Tamblyn, S., deGrosbois, J., Taylor, D. and Stratton, J. An outbreak of Escherichia coli O157:H7 infection associated with unpasteurized non-commercial, custom-pressed apple cider – Ontario, 1998. *Canada Communicable Disease Report* 1999; 25: 113–17.

Tuttle, J., Gomez, T., Doyle, M.P., Wells, J.G., Zhao, T. and Tauxe, R.V. *et al.* Lessons from a large outbreak of Escherichia coli O157:H7 infections: insights into the infectious dose and method of widespread contamination of hamburger patties. *Epidemiology & Infection* 1999; 122: 185–92.

Unpasteurised apple juice causes outbreak of haemolytic uraemic syndrome in North America. *Communicable Disease Report* 1996; *CDR Weekly* 6: 399.

Williams, R.C., Isaacs, S., Decou, M.L., Richardson, E.A., Buffett, M.C. and Slinger, R.W. *et al.* Illness outbreak associated with Escherichia coli O157:H7 in Genoa salami. E. coli O157:H7 Working Group. *CMAJ* 2000; 162: 1409–13.

Wright, J.R., Sumner, S.S., Hackney, C.R., Pierson, M.D. and Zoecklein, B.W. Efficacy of ultraviolet light for reducing Escherichia coli O157:H7 in unpasteurized apple cider. *Journal of Food Protection* 2000; 63: 563–7.

CONTROL POINT 12: CONSUMPTION OF READY-TO-EAT FOODS CROSS-CONTAMINATED BY VTEC O157

Aim: To prevent cross-contamination of VTEC O157 from raw foods to ready-to-eat foods

Agencies: FSA, local authorities, retail food and catering industry organisations.

Legislation

Food Safety Act 1990.
Food Safety (General Food Hygiene) Regulations 1995.
Food Safety (Temperature Control) Regulations 1995.
Meat Products (Hygiene) Regulations 1994.

Enforcement arrangements

The Food Safety Act 1990 makes it an offence to sell food for human consumption which fails to comply with food safety requirements. A food fails to comply with food safety requirements if it has been rendered injurious to health, is unfit for human consumption or is so contaminated that it would not be reasonable to expect it to be used for human consumption. The legislation is enforced by authorised officers (Environmental Health Officers and other authorised officers) working for a food authority (local authorities). Officers have the power to inspect any food intended for human consumption and to order the detention or seizure or food by giving notice to the person in charge of the food. Where food is seized, it is brought before a justice of the peace, who has the power to condemn the food and order that it be destroyed so as to prevent it from being used for human consumption.

The Meat Products (Hygiene) Regulations 1994 require premises not selling to the final consumer to be licensed.

The Food Safety (General Food Hygiene) Regulations 1995, are made under the Food Safety Act and implement Council Directive 93/43/EEC on the hygiene of foodstuffs. They apply to all stages of food production except primary production. The regulations require proprietors of food businesses to identify any step in the activities of the food business which is critical to ensuring food safety and ensure that adequate safety procedures are identified, implemented, maintained and reviewed on the basis of analysis of hazards; identification of points where hazards may occur;

deciding on critical points; identification and implementation of effective control and monitoring of those points and review.

There are also several requirements relating to the layout, design and construction of food premises which must be kept clean and in good repair and condition. Also, the requirements for adequate wash hand basins, lighting and drainage will all play a role in the prevention of cross-contamination.

Conveyances used for transporting foodstuffs must be clean and in good repair and containers must not be used for anything other than foods. Effective separation of products must be maintained to prevent contamination. Food handlers must maintain a high degree of personal cleanliness.

A person found guilty of contravening the above can incur a fine or imprisonment up to a term of 2 years. If, following a conviction, the court is satisfied that a 'health risk condition' is fulfilled, the court can prohibit the use of a process or treatment, equipment or premises. The 'health risk condition' is deemed to be fulfilled if any process, treatment, premises or equipment involves risk of injury to health.

Failure to comply with the regulations can be addressed via an improvement notice served on the proprietor specifying the measures which must be taken to secure compliance and a period within which they must be carried out.

Authorised officers can also serve emergency prohibition notices where the 'health risk condition' is fulfilled (see above) but where the risk of injury is imminent.

Compliance with the food hygiene regulations is assessed during routine food hygiene inspections. The frequency of these inspections is prescribed by a Code of Practice (Code of Practice No. 9 Food Hygiene Inspections). Individual premises are assigned an inspection rating based on a number of points. The premises with the highest number of points (category A) are inspected most frequently (at least every 6 months). Points are awarded for type and method of processing, consumers at risk, level of compliance and confidence in management. An additional score of 20 should be included where there is a significant risk of food being contaminated with *E.coli* O157 or other VTEC. Significant risk means a higher probability that an incident may occur and the following matters should be taken into account in assessing the risk:

(a) the potential for contamination/cross-contamination by the specified organism
(b) survival and growth of the specified organism
(c) the existence of hazard analysis systems and confidence in their implementation including documentation and records of monitoring of controls

(d) the extent and relevance of training undertaken by managers, supervisors and food handlers
(e) whether intervention by food authorities is necessary to reduce the probability of an incident occurring.

Between 1997 and 1998 additional funding of £19 million was made available for implementing the key recommendations of the Pennington Report, i.e. increasing the level of food safety enforcement in high risk premises and accelerating the implementation of food safety control systems based on HACCP.

In addition to the above requirements, certain food premises now require a licence issued by food authorities. Licensing requirements were introduced in England in November 2000 and in Wales in July 2001. Similar requirements were introduced in Scotland in October 2000. To obtain a licence, butchers and other food outlets selling open raw meat and ready-to-eat products from the same premises must have complied with the statutory hygiene requirements, operate a documented HACCP and enhanced food safety controls and staff hygiene training. A draft Industry Guide for butchers was issued for consultation in England in April 2001.

Figure 7.13 Control Point 12: Consumption of ready-to-eat foods; cross-contamination by VTEC O157

Published UK guidance/recommendations

ACMSF: We recommend that all relevant sectors of the food industry adopt a HACCP-based approach to prevent survival of or contamination by VTEC.

We recommend that procedures to ensure effective separation of raw and cooked meats in all premises should be determined by the application of hazard analysis.

Given the urgency of the situation, every effort should be made to implement and stringently enforce existing legislative requirements for hazard analysis and training. Hazard analysis should be documented, where appropriate.

Codes of Practice under section 40 of the Food Safety Act, which give guidance on efficient and effective enforcement of the hazard analysis requirement for the categories of premises A, B and C (Code of Practice 9), should be developed.

An Industry Guide to Good Hygiene Practice should be urgently developed under the Food Safety (General Food Hygiene) Regulations 1995 specifically for butchers and producers and retailers handling or preparing cooked meat and cooked meat products.

Appropriate training for food handlers, including supervisors and managers, is vital and should be accelerated by this sector of the industry.

Pennington: HACCP (i.e. the approach and all seven principles) should be adopted by all food businesses to ensure food safety. While this is being negotiated into European Union and domestic legislation, implementation and enforcement of the HACCP principles contained in the existing legislation should be accelerated.

The government should seek to have all the HACCP elements negotiated within the Horizontal Directive.

The government should review the application of the Meat Products (Hygiene) Regulations 1994, and the guidance issued subsequently, to clarify the position regarding which premises are intended to be covered by the regulations.

Pending HACCP implementation, selective licensing arrangements for premises not covered by the Meat Products (Hygiene) Regulations 1994 should be introduced by new regulations.

The licensing arrangements should include appropriate requirements for the documentation of hazard analysis, labelling and record-keeping to facilitate product recall and temperature control and monitoring. In relation to training, there should be a requirement for all food handlers to have undertaken at least basic food training and for all supervisory staff (and those who run small, one-person operations) to be trained to at least intermediate level. In addition the licence should cover matters relating to the suitability of premises, equipment and hygiene practices to a level equivalent to that required by the 1994 Regulations.

In relation to the physical separation requirements of licensing:

(a) There should be separation, in storage, production, sale and display, between raw meat and unwrapped cooked meat/meat products and other ready to eat foods. This should include the use of separate refrigerators and production equipment, utensils and, wherever possible, staff.
(b) Where the use of separate staff cannot be achieved, alternative standards (such as the completion and implementation by the operator of a HACCP or the provision and use of additional facilities, e.g. for hand washing in the serving area) might be regarded as sufficient to permit the award of a licence.
(c) Where neither (a) nor (b) above can be achieved, the premises concerned should not be permitted to sell both raw and unwrapped cooked meat/products (although they may be permitted to sell pre-wrapped cooked/ready to eat meat products prepared elsewhere and brought in for that purpose.

Food hygiene training should be provided wherever possible within the primary and secondary curriculum.

Guidance and education about food handling and hygiene should be included in all food and catering education and training courses and should be reinforced through periodic advertising and awareness initiatives.

Steps should be taken by local authorities to encourage the adoption of HACCP principles in non-registered premises where there is catering for functions for groups of people involving the serving of more than just tea, coffee and confectionery goods.

Employers should ensure that food handlers, in particular those working with vulnerable groups and/or in sensitive areas such as nursing homes and day-care centres, are aware of and implement good hygiene practice. They should be trained in food hygiene at least to the basic and preferably intermediate level.

The government should give a clear policy lead on the need for the enforcement of food safety measures and the accelerated implementation of HACCP.

The government and local authorities should ensure that there are available suitable and adequate Environmental Health Officer skills and resources to address enforcement and education/awareness issues.

The government should consider earmarking local authority funds for these purposes.

Local authorities should designate an Environmental Health Officer, with appropriate training, experience and expertise to head food safety within the authority.

Evidence presented to VTEC taskforce: Issues included marketing of infection free meat, new novel food processing methods to reduce bacterial load, risk reduction throughout the food chain, advice/discussion on risk assessment in controllable (e.g. food chain) circumstances, improved education/communication, use of media and pooling of expertise and experience.

WHO: For food retail and foodservices: The objective here should be to reduce the risk of direct and indirect cross-contamination between raw and ready-to-eat foods. This can be achieved through time and spatial physical separation, together with safe food hygiene practices. Judgements on the degree of separation and handling practices should be made within the framework of a HACCP based approach. In practice, consideration will need to be given to the manner in which separation will be achieved in storage, production, sale and display. It may involve separate refrigeration, work areas, equipment, utensils and staff.

Food handlers, including home caterers, should be trained in the principles of food hygiene and the application of HACCP, with particular attention to the special precautions of preparing and serving food to vulnerable groups, in accordance with WHO recommendations Hygiene in Food-Service and Mass Catering establishments (WHO/FNU/FOS/94.5); *Safe Food Handling – A Guide for Managers of Food Service Establishments*, by M. Jacob (WHO 1989).

Consumer education

Existing education programmes should be reviewed to ensure that they adequately address the low infectious dose of EHEC, the understanding of severe, long-term effects of infection, and the need for thorough cooking.

Food hygiene training should be carried out wherever possible within the primary and secondary school curriculum through the use of well-designed educational material, such as those provided by WHO Food, *Environment and Health – A Guide for Primary School Teachers*, by T. Williams, A. Moon and M. Williams (WHO 1990).

An education strategy on food safety for the consumer, conducted by a variety of organisations, including government, the food industry, consumer associations, trade associations, and consumer groups, should provide consistent, clearly understood messages, soundly based on science. Messages could be tested by focus groups. Conflicting messages must be avoided.

Educational programmes should use all means of disseminating information, including the media (newspaper, TV, radio) coverage, leaflets, lectures/lessons, training courses, labelling, in-store posters/leaflets.

The concept of risk communication involving all stakeholders (interested parties), including consumers, should be adopted to reduce EHEC infections.

Key points and practical issues

- The largest outbreak of VTEC O157 in the United Kingdom occurred in Scotland, in November 1996. Over 400 were identified and 20 people died. All of the clinical isolates were indistinguishable on PFGE fingerprinting. The majority of cases were reported in Lanarkshire although cases were also reported from Forth Valley, Greater Glasgow and Lothian. The epidemiological outbreak investigation comprised a cohort study of approximately 100 (mainly elderly) people who attended a church luncheon on 17 November 1996. Cases were also associated with a birthday party held on 23 November. The epidemic strain was isolated from gravy which was served at the above church luncheon. *E.coli* O157 was also isolated from a 6 lb piece of unopened vacuum-packed cooked beef, prepared by the suspect butcher. Secondary, person-to-person transmission was also suspected. The implicated butcher had a substantial wholesale and retail trade involving the production and distribution of raw and cooked meats and bakery products. The onsets of outbreak cases spanned a month and it was suspected that the latter ones were a result of person-to-person spread.
- Despite significant capital investment, the layout and design of the implicated butcher constrained measures to allow effective separation of cooked and raw products. The premises concerned was exempt from Meat Products (Hygiene) Regulations 1994 on the grounds that they were not handling meat products for sale other than to the final consumer.
- This outbreak and the subsequent report of the circumstances leading to the outbreak (Pennington) led to legislative changes in the UK concerning the production and handling of meat and meat products. In particular, the issue of cross-contamination in premises selling open raw meat and cooked meat products has been addressed. Many further outbreaks have been reported where a link with a particular food business but not with any particular food was established. It is likely that cross-contamination played a role in spreading the organism between food products. Cooked sliced meats and cooked sliced meat products (e.g. sandwiches) have been implicated in several further outbreaks. Studies of sporadic cases have also highlighted cooked meat products as a risk factor.
- The prevention of cross-contamination requires a meticulous approach to hygiene in food premises. Current legislation and guidance endorses a HACCP approach to hygiene. HACCP is based on the following seven principles.

1 Analysis of the potential food hazards in a food business operation.
2 Identification of the points in those operations where food hazards may occur.
3 Deciding which of the points identified are critical to ensuring food safety ('critical points').
4 Identification and implementation of effective control and monitoring procedures (including critical limits), at those critical points.
5 Verification to confirm that the Hazard Analysis and Critical Control Points system is working effectively.
6 Review of the analysis of food hazards, the critical points and the control and monitoring procedures periodically, and whenever the food business' operations change.
7 Documentation of all procedures appropriate to the effective application of the principles listed in (1) to (6).

Further reading

Jackson, L.A., Keene, W.E., McAnulty, J.M., Alexander, E.R., Diermayer, M., Davis, M.A., Hedberg, K., Boase, J., Barrett, T.J., Samadpour, M. and Fleming, D.W. Where's the beef? The role of cross-contamination in four chain restaurant-associated outbreaks of Escherichia coli O157:H7 in the Pacific Northwest. *Archives of Internal Medicine* 160 (15): 2380–5, 2000 Aug 14–28.

Parry, S.M., Salmon, R.L., Willshaw, G.A. and Cheasty, T. (1998) Risk factors for and prevention of sporadic infections with vero cytotoxin (shiga toxin) producing Escherichia coli O157. *The Lancet* 351 (9108): 1019–22.

CONTROL POINT 13: PERSON-TO-PERSON TRANSMISSION VIA INFECTED FOOD HANDLERS OR FAECAL–ORAL TRANSMISSION

Aim: To prevent VTEC O157 being transmitted by an infected individual via food or by a faecal–oral route

Agencies: FSA, HSE, local authorities, retail food and catering industry organisations.

Legislation

Food Safety Act 1990.
Food Safety (General Food Hygiene) Regulations 1995.

Health and Safety at Work etc. Act 1974.
Control of Substances Hazardous to Health Regulations 1999.

Enforcement arrangements

The control of infection in children's day care and in laboratories and other establishments where employees are exposed to human faeces is governed by the Health and Safety at Work etc. Act 1974 and the Control of Substances Hazardous to Health Regulations 1999. The Health and Safety Executive enforce the legislation on premises where the main purpose is education or health care (hospitals, nursing homes and laboratories) and in children's day-care facilities operated by local authorities. In residential homes and private children's day-care centres enforcement of this legislation is the responsibility of the local authority environmental health departments. Enforcement is via improvement or prohibition notices and fines. The COSHH Regulations 1999 require employers and self-employed people to assess the risks to health from work activities which involve a hazardous substance (e.g. a microorganism); prevent or, when this is not reasonably practicable, adequately control exposure to the hazardous substances; introduce and maintain control measures; inform, instruct and train employees about the risks and precautions to be taken; regularly review risk assessments and the effectiveness of control measures.

The Food Safety Act 1990, makes it an offence to sell food for human consumption which fails to comply with food safety requirements. A food fails to comply with food safety requirements if it has been rendered injurious to health, is unfit for human consumption or is so contaminated that it would not be reasonable to expect it to be used for human consumption. The legislation is enforced by authorised officers (Environmental Health Officers and other authorised officers) working for a food authority (local authorities). Officers have the power to inspect any food intended for human consumption and to order the detention or seizure or food by giving notice to the person in charge of the food. Where food is seized, it

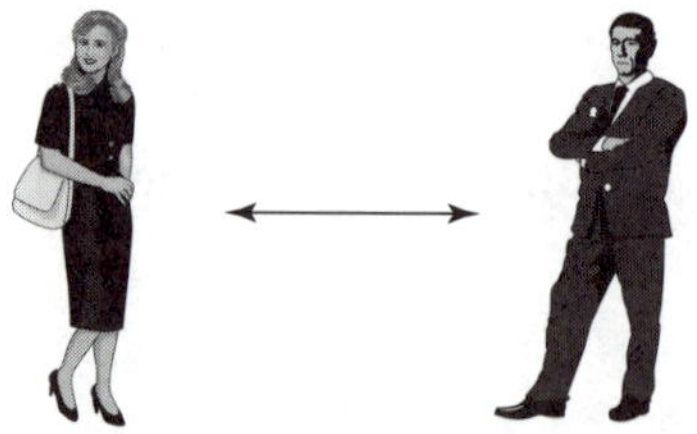

Figure 7.14 Control Point 13: Person-to-person transmission via infected food handlers or faecal–oral transmission

is brought before a justice of the peace, who has the power to condemn the food and order that it be destroyed so as to prevent it from being used for human consumption.

The Food Safety (General Food Hygiene) Regulations 1995, are made under the Food Safety Act and implement Council Directive 93/43/EEC on the hygiene of foodstuffs. They apply to all stages of food production except primary production. The regulations require proprietors of food businesses to identify any step in the activities of the food business which is critical to ensuring food safety and ensure that adequate safety procedures are identified, implemented, maintained and reviewed on the basis of analysis of hazards; identification of points where hazards may occur; deciding on critical points; identification and implementation of effective control and monitoring of those points and review.

The regulations also require persons working in a food handling area who know or suspect that they are carrying a disease which may be transmitted via food to report this to the proprietor of the food business. Persons who are known or suspected to be infected by a disease which is likely to be transmitted through food are not permitted to work in any food handling area in any capacity in which there is a likelihood of directly or indirectly contaminating food with pathogenic organisms.

Failure to comply with the regulations can be addressed via an improvement notice served on the proprietor specifying the measures which must be taken to secure compliance and a period within which they must be carried out.

Authorised officers can also serve emergency prohibition notices where the 'health risk condition' is fulfilled (see above) but where the risk of injury is imminent.

Compliance with the food hygiene regulations is assessed during routine food hygiene inspections. The frequency of these inspections is prescribed by a Code of Practice (Code of Practice No. 9 Food Hygiene Inspections). Individual premises are assigned an inspection rating based on a number of points. The premises with the highest number of points (category A) are inspected most frequently (at least every 6 months). Points are awarded for type and method of processing, consumers at risk, level of compliance and confidence in management. An additional score of 20 should be included where there is a significant risk of food being contaminated with *E.coli* O157 or other VTEC. Significant risk means a higher probability that an incident may occur and the following matters should be taken into account in assessing the risk:

(a) the potential for contamination/cross-contamination by the specified organism
(b) survival and growth of the specified organism
(c) the existence of hazard analysis systems and confidence in their imple-

mentation including documentation and records of monitoring of controls

(d) the extent and relevance of training undertaken by managers, supervisors and food handlers

(e) whether intervention by food authorities is necessary to reduce the probability of an incident occurring.

Finally, the powers and duties regarding reporting, investigation and control of infectious diseases in England and Wales are vested in local authorities by the Public Health (Control of Disease) Act 1984. VTEC O157, considered as 'food poisoning' and hence reportable under the 1984 Act, must be reported to the Proper Officer of the local authority who is empowered to take action to prevent further spread. In Wales, individual cases of food poisoning (and hence VTEC O157) are investigated by the environmental health department. The local authority (or Proper Officer) has the power to require the infected person to refrain from an occupation involving food, or, to refrain from attending school (if they are a child) or to take other measures to prevent the spread of infection (Schedule 4, Public Health (Infectious Diseases) Regulations 1988; Public Health (Control of Disease) Act 1984).

Guidance on the most appropriate policies of exclusions have been issued (see Further reading). The Subcommittee of the PHLS Advisory Committee on gastrointestinal Infections have issued detailed guidance on certain groups (whether they are cases or household contacts of known cases) posing a particular risk of spreading infection. These are food handlers, care staff, children under 5 and older children and adults who are unable to implement good standards of personal hygiene.

General guidance for managers in the food industry on infections which can be transmitted by infected staff is also available (*Food Handlers Fitness to Work – Guidelines for Food Business Managers*).

Published UK guidance/recommendations

ACMSF: We recommend that all relevant sectors of the food industry adopt a HACCP-based approach to prevent survival of or contamination by VTEC.

Pennington: Food hygiene training should be provided wherever possible within the primary and secondary curriculum.

Guidance and education about food handling and hygiene should be included in all food and catering education and training courses and should be reinforced through periodic advertising and awareness initiatives.

Steps should be taken by local authorities to encourage the adoption of HACCP principles in non-registered premises where there is catering for functions for groups of people involving the serving of more than just tea, coffee and confectionery goods.

Employers should ensure that food handlers, in particular those working with vulnerable groups and/or in sensitive areas such as nursing homes and day-care centres, are aware of and implement good hygiene practice. They should be trained in food hygiene at least to the basic and preferably intermediate level.

Evidence presented to VTEC taskforce: Person-to-person spread proposed practical steps.

Use gloves when washing soiled clothes; wash hands after using toilet; do not share towels; use disinfectant soap; all siblings under 7 years of age associated with outbreak to be routinely tested; separate staff for toiletry and food handling; stool charts within nurseries.

WHO: Food handlers, including home caterers, should be trained in the principles of food hygiene and the application of HACCP, with particular attention to the special precautions of preparing and serving food to vulnerable groups, in accordance with WHO recommendations Hygiene in Food-Service and Mass Catering Establishments (WHO/FNU/FOS/94.5); *Safe Food Handling – A Guide for Managers of Food Service Establishments*, by M. Jacob (WHO 1989).

Consumer education

Existing education programmes should be reviewed to ensure that they adequately address the low infectious dose of EHEC, the understanding of severe, long-term effects of infection, and the need for thorough cooking.

Food hygiene training should be carried out wherever possible within the primary and secondary school curriculum through the use of well-designed educational material, such as those provided by WHO Food, *Environment and Health – A Guide for Primary School Teachers*, by T. Williams, A. Moon and M. Williams (WHO 1990).

An education strategy on food safety for the consumer, conducted by a variety of organisations, including government, the food industry, consumer associations, trade associations, and consumer groups, should provide consistent, clearly understood messages, soundly based on science. Messages could be tested by focus groups. Conflicting messages must be avoided.

Educational programmes should use all means of disseminating information, including the media (newspaper, TV, radio) coverage, leaflets, lectures/lessons, training courses, labelling, in-store posters/leaflets.

The concept of risk communication involving all stakeholders (interested parties), including consumers, should be adopted to reduce EHEC infections.

Non-foodborne transmission

Staff in laboratories working with EHEC organisms should be made aware of the due care required because of their low infectious dose and severity of disease caused by these organisms.

Key points and practical issues

- VTEC O157 can be transmitted by an infected person who excretes the bacteria in faeces for up to several weeks following the onset of their symptoms. It is also possible to be infected by VTEC O157 and to excrete the organism without showing any signs of illness. Faecal-oral transmission can occur directly and via contaminated articles such as toilets and toys and infected food handlers who do not observe strict hygiene after visiting the toilet can introduce the bacteria into foods. VTEC O157 is readily transmitted via the faecal-oral route as it has a very low infectious dose. Several outbreaks have been reported where person-to-person transmission has been identified. These outbreaks are most common in institutional settings where individuals are unable to observe strict hygienic precautions during toileting by virtue of their age (very young children or the elderly) or a disability (residential care homes) or illness (hospitals). During outbreaks, staff working at these premises have also been infected. Outbreaks in children's day care are a particular problem as the incidence of VTEC O157 is highest in children under 5 and the complication of HUS is most likely to occur in this age group. In many instances it has been necessary to completely close the children's day-care centre in order to contain the outbreak.

Group 1	Food handlers whose work involves touching unwrapped foods to be consumed raw or without further cooking
Group 2	Health care, preschool nursery, or other staff who have direct contact or contact through serving food, with highly susceptible patients or people in whom an intestinal infection would have particularly serious consequences
Group 3	Children under 5 years of age attending nurseries, play groups, or other similar groups
Group 4	Older children and adults who are unable to implement good standards of personal hygiene – for example those with learning disabilities or special needs: and people in circumstances where hygienic arrangements may be unreliable – for example, temporary camps housing displaced persons. Under exceptional circumstances children in infant schools may be considered to fall into this group.

Figure 7.15 Groups that pose a special risk of spreading infection

Where such measures are taken it is vital to ensure that infection is not further spread within the community by children attending alternative organised day care. Communication between the investigating authorities and parents is vital.

- Cases of VTEC O157 are also a potential source of infection to family members and others living with them. Prompt advice on practical steps to prevent spread within the home is therefore vital for both cases identified during outbreaks and sporadic cases.
- In some outbreaks, it has been estimated that up to 10 per cent of the total identified cases were secondary cases who acquired the infection from an infected family member.
- Persons employed in occupations where they are required to come into contact with human faeces are also at risk. Several laboratory acquired infections have been reported and staff working with young children, in residential care and with sewage or wastewater are also at risk.
- Contact with a person with diarrhoea has been identified as a risk factor in several case control studies of sporadic cases.

Further reading

Advisory Committee on Dangerous Pathogens. *Supplement to: Categorisation of biological agents according to hazard and categories of containment (4th edition 1995).* Sudbury: HSE Books, 1998.

Al-Jadar, L., Salmon, R.L., Walker, A.M., Williams, H.M., Willshaw, G.A. and Cheasty, T. (1999) Outbreak of Escherichia coli O157 in a nursery: lessons for prevention. *Archives of Diseases in Childhood* 81: 60–63.

Expert Working Group convened by the Department of Health (1996) Guidelines for Food Business Managers. DOH.

Parry, S.M. and Salmon, R.L. (1998) Sporadic STEC O157 Infection: Secondary Household Transmission in Wales. *Emerging Infectious Diseases* 4 (4): 657–61.

Subcommittee of the PHLS Advisory Committee on gastrointestinal Infections. (2000) Guidelines for the control of infection with verocytotoxin producing Escherichia coli (VTEC). *Communicable Disease and Public Health* 3; 14–23.

Glossary

Adjustment Minimising, by statistical methods, the effects of differences in composition of the populations being compared.

Aetiology The cause of a disease.

Agglutination Clumping together of antigens by antibodies so that a visible agglutinate is formed.

Antibody Protein formed in direct response to the introduction into an individual of an antigen, which combines with its specific antigen, e.g. to neutralise toxins or destroy bacteria.

Antigen A substance that elicits an immune response when introduced into an individual.

Antiserum A solution that contains antibodies.

Association Statistical dependence between two or more variables. The presence of an association does not necessarily imply a causal relationship.

Attributable proportion The proportion by which the incidence rate of the outcome in the entire population would be reduced if the exposure were eliminated.

Bacteriophage or 'phage Bacteriocidal viruses.

Bias Deviation of results or inferences from the truth, or processes leading to such deviation.

Beefburger patty Minced beef (sometimes with added onion and flavouring) formed into flat round shapes and usually eaten in a bread bun. Sometimes referred to as a beefburger although the whole product, including the bun, is also referred to as a beefburger.

Bowel necrosis Death of parts of the wall of the bowel.

Bowel perforation A hole through the wall of the bowel.

Case control study An observational epidemiological study of persons with a disease and a suitable control group of persons without the disease. The diseased and non-diseased are compared with regard to how frequently a particular attribute is present in each of the groups.

Case fatality rate The proportion of cases of a specified condition which are fatal within a specified time.

Case series A group of cases identified by a particular study.

Catering premises Commercial food premises serving food for immediate consumption either on the premises (e.g. restaurant) or elsewhere (takeaway).

Clinical isolate Isolate from a human specimen.

Cohort A designated group of persons who are followed or traced over a period of time.

Colon The large intestine consisting of the ascending colon (first section), transverse colon (second section), descending colon (third section) and sigmoid colon (fourth section).

Cooked sliced meats Meats cooked and cooled and sold cold and sliced, e.g. sliced ham.

Confidence intervals The computed interval with a given probability (here it is 95 per cent) that the true value is contained within the interval.

Confounding A situation in which a measure of the effect of an exposure on risk is distorted because of exposure to other factor(s) that influence the outcome under study.

Cross-contamination Where organisms on one food (usually raw food) become transferred to other foods via the foodhandler, utensils, the environment, etc. and where the second foods do not receive treatment which would render them safe prior to consumption.

DNA hybridisation Matching a DNA fragment to a target DNA sequence.

D value The time required at a given temperature to reduce the number of viable cells of a given microorganism to 10 per cent of the initial number, usually quoted in minutes.

Environmental Health Officer Individuals with a qualification in Environmental Health which is recognised by the Chartered Institute of Environmental Health and who work in local government, the civil service or in the private sector controlling those aspects of the environment that affect public health.

Epidemic curve A graphic plotting of the distribution of cases by time of onset.

Epidemiology The study of the distribution and determinants of health-related states or events in specified populations, and the application of this study to the control of health problems.

Epithelial cells Cells that form the layer (epithelium) lining the inner surface of the intestines.

Endothelial cells Cells that form the layer (endothelium) lining the inner surface of blood and lymph vessels and the heart.

Enzyme linked immunosorbent assay A serological test that uses enzyme reactions as indicators of the presence/absence of an antibody/antigen reaction.

Escherichia coli Gram negative bacillus bacteria commonly found in the intestines of warm blooded animals including humans.

Exposure Proximity and/or contact with possible source of a disease agent in such a manner that effective transmission of the agent may occur.

Gastrointestinal tract stricture Narrowing of the gastrointestinal tract.

Gene probe A DNA fragment that has been labelled with a marker to indicate when DNA hybridisation has occurred.

Geographic clustering Aggregation of disease in space in amounts that are believed to be greater than could be expected by chance.

Haemolytic anaemia Condition where the destruction of red blood cells is about six times the normal rate and the supply of new cells from the bone marrow cannot meet the demand.

Haemorrhagic colitis Disease characterised by inflammation and frank blood from the colon.

Hazard An event with adverse effects on health.

Hazard analysis critical control point A systematic method of identifying (HACCP) potential hazards in food production processes, identifying the points at which control is critical and maintaining and monitoring suitable controls.

Hospital based study A study of subjects recruited because they have been referred to a particular hospital and may not, therefore, be representative of all such individuals in a population.

ID_{50} Number of organisms that cause symptomatic infection in 50 per cent of those exposed.

Immunomagnetic separation (IMS) A technique for isolating a particular organism using magnetic beads coated with antibodies to that organism.

Incidence The number of new diseases that occur in a defined period divided by the population at risk of experiencing the event during that period.

Incubation period The time interval between invasion by an infectious agent and the appearance of the first sign or symptom of the disease.

Index case The first case in a family or other defined group to come to the attention of the investigator.

Infectious dose The number of bacteria needed to cause an infection.

Isolate Bacterial growth originating from a particular sample.

Intussusception Where part of the gastrointestinal tract telescopes with the part beneath causing an obstruction and strangulation of the part which has been telescoped.

Irradiation The process of applying radiation energy to food to sterilise or preserve it by destroying microorganisms, parasites, insects or other pests. The type of radiation used is called ionising radiation because it produces ions. Sources of ionising radiation are gamma rays, electron beams and X rays.

Laparotomy Surgical operation to cut open the abdominal cavity.

Lipopolysaccharide (LPS) The part of the outer membrane of gram negative bacterial cells which functions as 'O' antigens.

Matching The process of making a study group and a comparison group comparable with respect to extraneous factors.

Mobile caterers Caterers cooking and serving food from mobile premises, e.g. 'burger vans'.

Multivariate analysis A set of techniques used when the variation of several variables has to be studied simultaneously.

Myocardial dysfunction Abnormal function of the heart muscle.

Normal distribution A distribution which has the following properties: a continuous, symmetrical distribution; both tails extend to infinity; the arithmetic mean, median and mode are identical; the shape is completely determined by the mean and standard deviation.

Odds ratio The ratio of the odds in favour of exposure among cases to the odds in favour of exposure among non-cases (controls).

'Phage See bacteriophage.

Point source outbreak An outbreak where a group of persons are exposed to a noxious influence that is common to the individuals in the group and where the exposure is brief and essentially simultaneous and the resultant cases all develop within one incubation period of the disease.

Polymerase chain reaction A technique that enables multiple copies of a DNA fragment to be generated by amplification of target DNA sequence.

Population based study A study based on subjects drawn from a definable population. The population is the denominator and/or sampling frame.

Potable water Drinking water.

Presumptive VTEC O157 An isolation of VTEC O157 made by a local microbiology laboratory which has not yet been confirmed as an *E.coli* O157 not as a VT producing strain by a Reference Laboratory.

Prevalence The number of events in a given population at a designated time.

Primary case The individual who introduces the disease into a family or group under study.

Prodrome Early symptom of a disease.

Prospective study A study where data collection continues with the passage of time.

Relative risk The ratio of the risk of disease among the exposed to the risk among the unexposed.

Reference Laboratory For *E.coli* O157 England and Wales and Northern Ireland this is PHLS LEP, Colindale (see contact details section). For

E.coli O157 isolated in Scotland, this is Scottish *E.coli* O157 Reference Laboratory at the Department of Medical Microbiology, Aberdeen.

Reservoir The natural habitat of the infectious agent.

Retail premises Commercial food premises where foods are sold for consumption elsewhere and not necessarily for immediate consumption.

Retrospective study A study where data is from past events.

Risk The likelihood of a hazard occurring.

Risk assessment A method of assessing how likely a hazard is to occur.

Secondary case Cases of an infection that occur among contacts within the incubation period following exposure to a primary case.

Sequelae Conditions that follow on from an earlier disease.

Serology The study of antigen-antibody reactions *in vitro*.

Serotyping A method of distinguishing varieties of bacteria (serotypes) by defining their antigenic properties on the basis of their reaction to known antisera.

Serum antibodies Antibodies found in the fluid fraction of coagulated blood.

Sorbitol MacConkey Agar A selective medium for the detection of *Escherichia coli* O157.

Source The person, animal, object or substance from which an infectious agent passes to a host. It should be clearly distinguished from the source of contamination which could be an entirely separate entity.

Standard deviation The positive square root of the variance. A summary of how widely dispersed the values are around the mean.

Statistical significance Usually expressed as a p value this is the probability of observing the study results under the null hypothesis.

Strain discrimination Any technique for distinguishing between species or sub-species or strains within sub-species.

Stratification Separating a sample into several sub-samples according to specified criteria.

Surveillance The ongoing and systematic collection, analysis and interpretation of health data in the process of describing and monitoring a health event.

Temporal clustering Aggregation of disease in time in amounts that are believed to be greater than could be expected by chance.

Thrombotic thrombocytopaenic purpura A condition resulting from the aggregation of platelets in various organs, characterised by fever with skin and central nervous system involvement, anaemia and kidney failure.

Titre A measurement of the quantity of antibodies in a serum.

Toxin A poisonous substance produced by a microorganism.

Variance The sum of squares of deviation from the mean of a set of values divided by the number of degrees of freedom.

Vehicle of infection The means whereby an infectious agent gets from its reservoir to a susceptible host. Examples include faecal–oral, respiratory and foodborne.

Virulence mechanism The way in which a pathogen causes disease.

Water activity A measure of the available water of a substance.

Zoonosis An infection transmissible under natural conditions from animals to humans.

Useful contact telephone numbers, addresses and websites

Centers for Disease Control and Prevention
1600 Clifton Road
Atlanta
GA 30333
USA
Tel: (00 404) 639 3311
Fax:
www.cdc.gov

Communicable Disease Surveillance Centre
61 Colindale Avenue
London
NW9 5EQ
Tel: 020 8200 6868
Fax: 020 8200 7868
http://www.phls.co.uk

Communicable Disease Surveillance Centre (Wales)
Abton House
Wedal Road
Roath
Cardiff
CF14 3QX
Tel: 029 20 521997
Fax: 029 20 521987
http://www.phls.wales.nhs.uk

Dairy Hygiene Inspectorate
Riverside Chambers
Castle Street
Taunton
TA1 4AP
Tel: 01823 285540

Fax: 01823 285594
http://www.foodstandards.gov.uk/

Department of Health
Richmond House
79 Whitehall
London
SW1A 2NS
Tel: 020 7972 2000
Fax: 020 7972 5758
Email: dhmail@doh.gsi.gov.uk
http://www.doh.gov.uk/

Drinking Water Inspectorate
Floor 2/A1
Ashdown House
123 Victoria Street
London
SW1E 6DE
Tel: 020 7944 5956
Fax: 020 7944 5969
Email: dwi_enquiries@detr.gov.uk
http://dwi.detr.gov.uk/

Environment Agency
Rio House
Waterside Drive
Aztec West
Almondsbury
Bristol
BS12 4UD
Tel: 01454 624400
Fax: 01454 624409
Email: enquiries@environment-agency.gov.uk
http://www.environment-agency.gov.uk

Food Standards Agency (England)
Room 245 Aviation House
125 Kingsway
London
WC2B 6NH
Tel: 020 7276 8000
Fax: 020 7238 6330
Email: helpline@foodstandards.gsi.gov.uk
http://www.foodstandards.gov.uk

Food Standards Agency (Scotland)
St Magnus House
6th Floor
25 Guild Street
Aberdeen
AB11 6NJ
Tel: 01224 285143
Fax: 01224 285168

Food Standards Agency (Wales)
1st Floor
Southgate House
Wood Street
Cardiff
CF10 1EW
Tel: 029 2067 8999
Fax: 029 2067 8918/8919

Food Standards Agency (Northern Ireland)
10B and 10C Clarendon Quay
Clarendon Dock
Clarendon Road
Belfast
BT1 3BW
Tel: 028 9041 7700
Fax: 028 9041 7726

Health and Safety Executive (England)
St Dunstans House
201 – 211 Borough High Street
London
SE1 1GZ
Tel: 020 7556 2100
Fax: 020 7556 2200
Email: hseinformationservices@natbrit.com
http://www.hse.gov.uk

Health and Safety Executive (Scotland)
Belford House
59 Belford Road
Edinburgh
EH4 3UE
Tel: 0131 247 2000
Fax: 0131 247 2121

Health and Safety Executive (Wales)
Government Buildings
Phase 1
Ty Glas
Llanishen
Cardiff
CF14 5SH
Tel: 029 20 263000
Fax: 029 20 263120

Ministry of Agriculture, Fisheries and Food (MAFF)
Nobel House
17 Smith Square
London
SW1P 3JR
Tel: 020 7238 3000
Fax: 020 7238 6591
Email: golondon.gol@go-regions.gov/uk
http://www.maff.gov.uk

Meat Hygiene Division
Food Standards Agency
P O Box 31037
Room 520
Ergon House
17 Smith Square
London
SW1P 3WG
Tel: 020 7238 6735
Fax: 020 7238 6138

Public Health Laboratory Service
Headquarters
61 Colindale Avenue
London
NW9 5DF
Tel: 020 8200 1295
Fax: 020 8200 8131
http://www.phls.co.uk/

Scottish Centre for Infection and Environmental Health
Clifton House
Clifton Place
Glasgow
G3 7LN
Tel: 0141 300 1100
Fax: 0141 300 1170
http://www.show.scot.nhs.uk/scieh

Scottish E.coli O157 Reference Laboratory
Lothian University Hospitals NHS Trust
Western General Hospital
Department of Clinical Microbiology
Crew Road
Edinburgh
EH4 2XU
Tel: 0131 537 1940
Fax: 0131 537 1024

Veterinary Laboratory Agency
New Haw
Addlestone
Surrey
KT15 3NB
Tel: 01932 341111
Fax: 01932 347046
Email:
http://www.gov.uk/vla

World Health Organization (WHO)
Headquarters
Avenue Appia 20
1211 Geneva 27
Switzerland
Tel: (+00 41 22) 791 21 11
Fax: (+00 41 22) 791 31 11
Email: info@who.int
http://www.who.int/

Bibliography

A report of the Study of Infectious Intestinal Disease in England. Food Standards Agency. HMSO London 2000.

Abdul-Raouf, U.M., Beuchat, L.R. and Ammar, M.S. (1993) Survival and growth of *Escherichia coli* O157:H7 in ground, roasted beef as affected by pH, acidulants, and temperature. *Applied and Environmental Microbiology* 59, 2364–8.

Adak, G.K., O'Brien, S.J., Gilliam, C. and Smith, H.R. (2000) The PHLS case-control study of Escherichia coli infection in England. *SCIEH Weekly Report Supplement* 34(5), 11.

Advisory Committee on the Microbiological Safety of Food (1995) Report on Verocytotoxin-Producing *Escherichia coli*. HMSO London.

Ahmed, R., Bopp, C., Borczyk, A. and Kasatiya, S. (1987) 'Phage-typing scheme for *Escherichia coli* O157:H7. *The Journal of Infectious Diseases* 155, 806–9.

Allaby, M.A.K. and Mayon-White, R. (1995) *Escherichia coli* O157: outbreak in a day nursery. *Communicable Disease Report Review* 5, R4–R6.

Anonymous (1994) *E.coli* O157 in Cattle. *Communicable Diseases Scotland Weekly Report* 28(40), 1.

Anonymous (1996) National outbreak of *Escherichia coli* O157 infection. *SCIEH Weekly Report* 30(50), 269.

Bahn, M.K., Raj, P., Levine, M.M., Kaper, J.B., Bhandari, N., Srivastava, R. and Sazawal, S. (1989) Enteroaggregative *Escherichia coli* associated with persistent diarrhea in a cohort of rural children in India. *The Journal of Infectious Diseases* 159, 1061–4.

Barrett, T.J., Green, J.H., Griffin, P.M., Pavia, A.T., Ostroff, S.M. and Wachsmuth, I.K. (1991) Enzyme-linked immunosorbent assays for detecting antibodies to Shiga-like toxin-I, shiga-like toxin II, and *Escherichia coli* O157:H7 lipopolysaccharide in human serum. *Current Microbiology* 23, 189–95.

Baqui, A.H., Sack, R.B., Black, R.E., Haider, K., Hossain, A., Alim, A.R.M.A., Yunus, M., Chowdhury, H.R. and Siddique, A.K. (1992) Enteropathogens associated with acute and persistent diarrhea in Bangladeshi children <5 years of age. *The Journal of Infectious Diseases* 166, 792–6.

Beaglehole, R., Bonita, R. and Kjellstrom, T. (1993) *Basic Epidemiology*. WHO, Geneva 1993.

Belongia, E.A., Osterholm, M.T., Soler, J.T., Ammend, D.A., Braun, J.E. and MacDonald, K.L. (1993) Transmission of *Escherichia coli* O157:H7 infection in

Minnesota child day-care facilities. *The Journal of the American Medical Association* 26, 883–8.

Besser, R.E., Lett, S.M., Weber, J.T., Doyle, M.P., Barrett, T.J., Wells, J.G. and Griffin, P.M. (1993) An outbreak of diarrhea and hemolytic uremic syndrome from *Escherichia coli* O157:H7 in fresh-pressed apple cider. *The Journal of the American Medical Association* 269, 2217–20.

Bilge, S.S., Clausen, C.R. and Lau, W. (1989) Molecular characterization of a fimbrial adhesin, F1845, mediating diffuse adherence of diarrhea-associated *Escherichia coli* to HEp-2 cells. *Journal of Bacteriology* 171, 4281–9.

Bitzan, M., Ludwig, K., Klemt, M., Konig, H., Buren, J. and Muller-Wiefel, D.E. (1993) The role of *Escherichia coli* O157 infections in the classical (enteropathogenic) haemolytic uraemic syndrome: results of a Central European, multicentre study. *Epidemiology and Infection* 110, 183–96.

Bryant, H.E., Athar, M.A. and Pai, C.H. (1989) Risk factors for *Escherichia coli* O157:H7 infection in an urban community. *The Journal of Infectious Diseases* 160, 858–64.

Carter, A.O., Borczyk, A.A., Carlson, J.A.K., Harvey, B., Hockin, J.C., Karmali, M.A., Krishnan, C., Korn, D.A. and Lior, H. (1987) A severe outbreak of *Escherichia coli* O157:H7-associated hemorrhagic colitis in a nursing home. *The New England Journal of Medicine* 317, 1496–1500.

Cebula, T.A., Payne, W.L. and Feng, P. (1994) MAMA/multiplex PCR: a specific means for shiga-like toxin typing and identification of the O157:H7 serotype. In: Karmali, M.A. and Goglio, A.G. (1994) Recent advances in Verocytotoxin-producing *Escherichia coli* infections. International Congress Series 1072. Excerpta Medica. Elsevier, Amsterdam, Lausanne, New York, Oxford, Shannon, Tokyo.

CDC (1993) From the Centers for Disease Control and Prevention. Update: Multistate Outbreak of *Escherichia coli* O157:H7 Infections from Hamburgers – Western United States, 1992–1993. *The Journal of the American Medical Association* 269, 2194–6.

CDC (1994) *Escherichia coli* O157.H7 outbreak linked to home-cooked hamburger – California, July 1993. *MMWR* 43, 213–15.

CDC (1995a) Enhanced detection of sporadic *Escherichia coli* O157:H7 infections – New Jersey, July 1994. Editorial note. *MMWR* 44, 418.

CDC (1995b) *Escherichia coli* O157:H7 outbreak linked to commercially distributed dry-cured salami – Washington and California, 1994. *MMWR* 44, 157.

CDSC (1996) VTEC O157 infection in West Yorkshire associated with the consumption of raw milk. *Communicable Disease Report Weekly* 6(21), 181.

Chapman, P.A., Wright, D.J. and Higgins, R. (1993) Untreated milk as a source of verotoxigenic *Escherichia coli* O157. *Veterinary Record* 133, 171–2.

Cieslak, P.R., Barrett, T.J., Griffin, P.M., Gensheimer, K.J., Beckett, G., Buffington, J. and Smith, M.G. (1993) *Escherichia coli* O157 infection from a manured garden [Letter]. *The Lancet* 342, 367.

Coia, J.E., Curnow, J., Reilly, W.J., Synge, B.F. and Sharp, J.C.M. (1995) Ten years' experience of *Escherichia coli* O157 infection in Scotland (1984–1993). *SCIEH Weekly Report* 29(1), 3–4.

Coia, J.E. and Hanson, M.F. (2000) Meat and Dairy Products Surveillance in South East Scotland. Proceedings from Zoonotic Infections in Livestock and the Risk to Public Health VTEC O157, Royal College of Physicians of Edinburgh (28th June).

Curnow, J. (1994) *E.coli* O157 Phage type 28 Infections in Grampian. *Communicable Diseases Scotland Weekly Report* 28(46), 1.

D'Aoust, J.Y., Park, C.E., Szabo, R.A., Todd, E.C.D., Emmons, D.B. and McKellar, R.C. (1988) Thermal inactivation of *Campylobacter* species, *Yersinia enterocolitica* and *hemorrhagic Escherichia coli* O157:H7 in fluid milk. *Journal of Dairy Science* 71, 3230–63.

Davis, B.S. and Brogan, R.T. (1994) Butchers, burgers and *E.coli* O157. *SCIEH Weekly Report* 28(43), 3–6.

Davis, B.S. and Brogan, R.T. (1995) A widespread community outbreak of *E coli* O157 infection in Scotland. *Public Health* 109, 381–8.

Day, N.P., Scotland, S.M., Cheasty, T. and Rowe, B. (1983) *Escherichia coli* O157:H7 associated with human infections in the United Kingdom. *The Lancet* 1, 825.

DOH (1991) Sir Donald Acheson's advice on cooking beefburgers. DOH Press Notice 14th February, 1991.

Dev, V.J., Main, M. and Gould, I. (1991) Waterborne outbreak of *Escherichia coli* O157 [letter]. *The Lancet* 337, 1412.

Dev, V.J., Main, M., Gould, I.M. and Sharp, J.C.M. (1991a) Still waters run deep. A possible waterborne outbreak of *Escherichia coli* O157. *Communicable Diseases Scotland Weekly Report* 25(41), 5–6.

Donnenberg, M.S. and Kaper, J.B. (1992) Enteropathogenic *Escherichia coli Infection and Immunity* 60, 3953–61.

Dytoc, M., Soni, R., Cockerill, F., de Azavedo, J., Louie, M., Brunton, J. and Sherman, P. (1993) Multiple determinants of verotoxin-producing *Escherichia coli* O157:H7 attachment-effacement. *Infection and Immunity* 61, 3382–91.

Elder, R.O., Keen, J.E., Siragusa, G.R., Barkocy-Gallagher, A.B., Koohmarale, M. and Laegreid, W.W. (2000) Correlation of enterohemorrhagic Escherichia coli O157 prevalence in faeces, hides and carcasses of beef cattle during processing. *PNAS* 97 (7) 2999–3003 (www.pnas.org).

Erickson, J.P., Stamer, J.W., Hayes, M., McKenna, D.N. and Van Alstine, L.A. (1995) An assessment of *Escherichia coli* O157:H7 contamination risks in commercial mayonnaise from pasteurised eggs and environmental sources, and behaviour in low-pH dressings. *Journal of Food Protection* 58(10), 1059–64.

Frost, J.A., Smith, H.R., Willshaw, G.A., Scotland, S.M., Gross, R.J. and Rowe, B. (1989) 'Phage-typing of Verocytoxin (VT) producing *Escherichia coli* isolated in the United Kingdom. *Epidemiology and Infection* 103, 73–81.

Frost, J.A., Cheasty, T., Thomas, A. and Rowe, B. (1993) 'Phage-typing of Verocytotoxin-producing *Escherichia coli* O157 isolated in the United Kingdom: 1989–1991. *Epidemiology and Infection* 110, 469–75.

Fukushima, H., Hashizume, T., Morita, Y., Tanaka, J., Azuma, K., Mizumoto, Y., Kaneno, M., Matsuura, M., Konma, K. and Kitna, T. (1999) Clinical experiences in Sakai City Hospital during the massive outbreak of enterohemorrhagic Escherichia coli O157 in Sakai City, 1996. Pediatrics International 41(2), 213–17.

Gammie, A.J., Mortimer, P.R., Hatch, L., Brierley, A.F.M., Chada, N., Walters, J.B. (1996) Outbreak of verocytotoxin-producing *Escherichia coli* O157 associated with cooked ham from a single source. *PHLS Microbiology Digest* 13(3), 142–5.

Garland Miller, L. and Kaspar, C.W. (1994) *Escherichia coli* O157:H7 acid tolerance and survival in apple cider. *Journal of Food Protection* 57(6), 460–4.

Glass, K.A., Loeffelholz, J.M., Ford, J.P. and Doyle, M.P. (1992) Fate of *Escherichia coli* O157:H7 as affected by pH or sodium chloride and in fermented dry sausage. *Applied and Environmental Microbiology* 58(8), 2513–16.

Griffin, P.M. (1995) *Escherichia coli* O157:H7 and other enterohemorrhagic *Escherichia coli*. In Blaser, M.J., Smith, P.D., Ravdin, J.I., Greenberg, H.B., Guerrant, R.L. eds. Infections of the Gastrointestinal Tract. Raven Press. New York 1995, 1–11.

Guerrant, R.L. and Thielman, N.M. (1995) Types of *Escherichia coli* Enteropathogens. In Blaser, M.J., Smith, P.D., Ravdin, J,I., Greenberg, H.B. and Guerrant, R.L. eds. Infections of the Gastrointestinal Tract. Raven Press. New York 1995, 37–51.

Hale, T.L., Sansonetti, P.J., Schad, P.P., Austin, S. and Formal, S.B. (1983) Characterization of Virulence Plasmids and Plasmid-Associated Outer Membrane Proteins in *Shigella flexneri*, *Shigella sonnei* and *Escherichia coli*. *Infection and Immunity* 40(1), 340–50.

Harris, J.R., Wachsmuth, I.K., Davis, B.R. and Cohen, M.L. (1982) High-molecular weight plasmid correlates with *Escherichia coli* enteroinvasiveness. *Infection and Immunity* 37, 1295–8.

Johnson, W.M., Lior, H. and Bezanson, G.S. (1983) Cytotoxic *Escherichia coli* O157:H7 associated with haemorrhagic colitis in Canada. *The Lancet* 1, 76.

Jones, I.G. and Rowarth, M. (1996) An outbreak of *Escherichia coli* O157 and Campylobacteriosis associated with contamination of a drinking water supply. *Public Health* 110, 277–82.

Karch, H., Heesemann, J., Laufs, R., O'Brien, A.D., Tacket, C.O. and Levine, M.M. (1987) A plasmid of enterohemorrhagic *Escherichia coli* O157:H7 is required for expression of a new fimbrial antigen and for adhesion to epithelial cells. *Infection and Immunity* 55, 455–61.

Karch, H., Russman, H., Schmidt, H., Schwarzkopf, A. and Heesemann, J. (1995) Long-term shedding and clonal turnover of enterohemorrhagic *Escherichia coli* O157 in diarrhoeal disease. *Journal of Clinical Microbiology* 33, 1602–5.

Karmali, M.A., Petric, M., Lim, C., Fleming, P.C., Arbus, G.S. and Lior, H. (1985) The association between idiopathic hemolytic uremic syndrome and infection by Verotoxin-producing *Escherichia coli*. *The Journal of Infectious Diseases* 151, 775–82.

Kelly, J., Oryshak, A., Wenetsek, M., Grabiec, J. and Handy, S (1990) The colonic pathology of *Escherichia coli* O157:H7 infection. *American Journal of Surgical Pathology*, 14(1), 87–92.

Khakhria, R., Duck, D. and Lior, H. (1990) Extended 'phage-typing scheme for *Escherichia coli* O157:H7. *Epidemiology and Infection* 105, 511–20.

Kohli, H.S., Chaudhuri, A.K.R., Todd, W.T.A., Mitchell, A.A.B. and Liddell, K.G. (1993) The Hartwood Hill Hospital *E.coli* O157 outbreak. *Communicable Diseases Scotland Weekly Report* 27(12), 8–11.

Levine, M.M. (1987) *Escherichia coli* that Cause Diarrhea: Enterotoxigenic, Enteropathogenic, Enteroinvasive, Enterohemorrhagic, and Enteroadherent. *The Journal of Infectious Diseases* 155, 377–89.

Levine, M.M., Kaper, J.B., Black, R.E. and Clements, M.L. (1983) New knowledge on pathogenesis of bacterial enteric infections as applied to vaccine development. *Microbiological Reviews* 47, 510–50.

Levine, M.M. and Edelman, R. (1984) Enteropathogenic *Escherichia coli* of classical serotypes associated with infant diarrhoea – epidemiology and pathogenesis. *Epidemiological Reviews* 6, 31–51.

Little, C.L., Mitchell, R.T. (2000) LACOTS/PHLS Co-ordinated Food Liaison Group Surveys: The Microbiological Examination of Ready-to-Eat Burgers Sampled Anonymously at the Point of Sale. Local Authorities Co-ordinated Body on Food and Training Standards and the Public Health Laboratory Service.

Locking, M. (2000) A case-control study of sporadic cases of Escherichia coli O157 infection in Scotland. SCIEH Weekly Report Supplement, 34(5): 9.

Louie, M., de Azavedo, J.C.S., Handelsman, M.Y.C., Clark, C.G., Ally, B., Dytoc, M., Sherman, P. and Bruton, J. (1993) Expression and characterisation of the eaeA gene product of the *Escherichia coli* serotype O157:H7. *Infection and Immunity* 61, 4085–92.

MacDonald, K.L., O'Leary, M.J., Cohen, M.L., Norris, P., Wells, J.G., Noll, E., Kobayashi, J.M. and Nlake, P.A. (1988) *Escherichia coli* O157;H7, an emerging gastrointestinal pathogen. Results of a one-year prospective, population-based study. *The Journal of the American Medical Association* 259, 3567–70.

Marier, R., Wells, J.G., Swanson, R.C., Callahan, W. and Mehlman, I.J (1973) An outbreak of enteropathogenic *Escherichia coli* foodborne disease traced to imported French cheese. *The Lancet* 2, 1376–8.

Marsh, J., MacLeod, A.F., Hanson, M.F., Emmanuel, F.X.S., Frost, J.A. and Thomas, A. (1992) A restaurant-associated outbreak of *E.coli* O157 infection. *Journal of Public Health Medicine* 14(1), 78–83.

Martin, M.L., Shipman, L.D., Wells, J.G., Potter, M.E., Hedberg, K., Wachsmuth, I.K., Tauxe, R.V., Davis, J.P., Arnoldi, J. and Tilleli, J. (1986) Isolation of *Escherichia coli* O157:H7 from dairy cattle associated with two cases of haemolytic uraemic syndrome [Letter]. *The Lancet* 2, 1043.

Mead, P.S., Finelli, L., Lambert-Fair, M.A., Champ, D., Townes, J., Hutwagner, L., Barrett, T., Spitalny, K. and Mintz, E. (1997) Risk factors for sporadic infection with *Escherichia coli* O157:H7. *Archives of Internal Medicine* 157, 204–8.

Moon, H.W., Whipp, S.C., Argenzio, R.A., Levine, M.M. and Gianella, R.A. (1983) Attatching and effacing activities of rabbit and human enteropathogenic *Escherichia coli* in pig and rabbit intestines. *Infection and Immunity* 41, 1340–51.

Morgan, D., Newman, C.P., Hutchinson, D.N., Walker, A.M., Rowe, B. and Majid, F. (1993) Verotoxin producing *Escherichia coli* O157 infections associated with the consumption of yoghurt. *Epidemiology and Infection* 111, 181–7.

Morgan, G.M., Newman, C., Palmer, S.R., Allen, J.B., Shepherd, W., Rampling, A.M., Warren, R.E., Gross, R.J., Scotland, S.M. and Smith, H.R. (1998) First recognised community outbreak of haemurrhagic colitis due to verotoxin-producing *Escherichia coli* O157: H7 in the UK. *Epidemiology and Infection* 101, 83–91.

Murano, E.A and Pierson, M.D. (1992) Effect of heat shock and growth atmosphere on the heat resistance of *Escherichia coli* O157:H7. *Journal of Food Protection* 55(3), 171–5.

O'Brien, A.D., Lively, T.A., Chen, M.E., Rothman, S.W. and Formal, S.B. (1983) *Escherichia coli* O157:H7 strains associated with haemorrhagic colitis in the United States produce a *Shigella dysenteriae* 1 (Shiga) like cytotoxin. [Letter] *The Lancet* 1, 702.

O'Donohue, E. (1996) The first reported outbreak of *E.coli* O157 in Ireland. *Infectious Diseases Bulletin* 4(2), 1–2.

Oelschlaeger, T.A., Barrett, T.J. and Kopecko, D.J. (1994) Some structures and processes of human epithelial involved in uptake of enterohemorrhagic *Escherichia coli* O157:H7 strains. *Infection and Immunity* 62, 5142–50.

Pai, C.H., Gordon, R., Sims, H.V. and Bryan, L.E. (1984) Sporadic cases of hemorrhagic colitis associated with *Escherichia coli* O157:H7. Clinical, epidemiological and bacteriological features. *Annals of Internal Medicine* 101(6), 738–42.

Pai, C.H., Ahmed, N., Lior, H., Johnson, W.M., Sims, H.V. and Woods, D.E. (1988) Epidemiology of sporadic diarrhea due to verocytotoxin-producing *Escherichia coli*: a two-year prospective study. *The Journal of Infectious Disease* 157, 1054–7.

Paiba, G.A. (2000) Prevalence of Verocytoxigenic *E.coli* O157 (VTEC O157) in Cattle, Sheep and Pigs at Slaughter in Great Britain and in Cattle on Farms in England and Wales. Proceedings from Zoonotic Infections in Livestock and the Risk to Public Health VTEC O157, Royal College of Physicians of Edinburgh (28th June).

Palmer, S.R. and Galbraith, N.S. (1998) The epidemiology of bacterial infections. In Hausler, W.J. and Sussman, M. (eds) Bacterial Infections. Volume 3. Topley and Wilsons Microbiology and Microbial Infections. Arnold. London 1998.

(The) Pennington Group (1997) Report on the circumstances leading to the 1996 outbreak of infections with *E coli* O157 in central Scotland. The implications for food safety and the lessons to be learned. The Stationery Office, Edinburgh 1997.

Pritchard, G.C., Willshaw, G.A., Bailey, J.R., Carson, T., Cheasty, T. Verocytotoxin-producing Escherichia coli O157 on a farm open to the public: outbreak investigation and longitudinal bacteriological study (2000). *Veterinary Record* 147 (10), 259–64.

Public Health Laboratory Service. Standard Operating Procedure. Investigation of faeces specimens for bacterial pathogens. B.SOP30. London, PHLS 1998.

Rahn, K., Renwick, S.A., Johnson, R.P., Wilson, J.C., Clarke, R.C., Alves, D., McEwen, S., Lior, H. and Spika, J. (1997) Persistence of Escherichia coli O157: H7 in dairy cattle and the dairy farm environment. *Epidemiology and Infection* 119 (2), 251–9.

Remis, R.S., MacDonald, K.L., Riley, L.W., Puhr, N.D., Wells, J.G., Davis, B.R., Blake, P.A. and Cohen, M.L. (1984) Sporadic cases of hemorrhagic colitis associated with *Escherichia coli* O157:H7. *Annals of Internal Medicine* 101, 624–6.

Renwick, S.A., Wilso, J.B., Clarke, R.C., Lior, H., Borczyk, A.A., Spika, J., Rahn, K., McFadden, K., Brouwer, A., Copps, A., Anderson, N.G., Alves, D. and Karmali, M.A. (1993) Evidence of direct transmission of *Escherichia coli* O157:H7 infection between calves and a human. *The Journal of Infectious Diseases* 168, 792–3.

Rice, D.H., Hancock, D.D., Vetter, R.L. and Besser, T.E. (1996) *Escherichia coli* O157 infection in a human linked to exposure to infected livestock. *The Veterinary Record* 136, 311.

Riley, L.W., Remis, R.S., Helgerson, S.D., McGee, H.B., Wells, J., Davis, B.R., Hebert, R.J., Olcott, E.S., Johnson, L.M., Hargrett, N.T., Blake, P.A. and Cohen, M.L. (1983) Hemorrhagic colitis associated with a rare *Escherichia coli* serotype. *The New England Journal of Medicine* 308, 681–5.

Roberts, C.L., Mshar, P.A., Cartter, M.L., Hadler, J.L., Sosin, D.M., Hayes, P.S. and Barrett, T.J. (1995) The role of heightened surveillance in an outbreak of *Escherichia coli* O157.H7. *Epidemiology and Infection* 115, 447–54.

Ryan, C.A., Tauxe, R.V., Hosek, G.W., Wells, J.G., Stoesz, P.A., McFadden, H.W. Jr., Smith, P.W., Wright, G.F. and Blake, P.A. (1986) *Escherichia coli* O157:H7 diarrhea in a nursing home: clinical, epidemiological, and pathological findings. *The Journal of Infectious Diseases* 154, 631–8.

Salmon, R.L., Farrell, I.D., Hutchinson, J.G.P., Coleman, D.J., Gross, R.J., Fry, N.K., Rowe, B. and Palmer, S.R. (1989) A christening party outbreak of haemorrhagic colitis and haemolytic uraemic syndrome associated with *Escherichia coli* O 157.H7. *Epidemiology and Infection* 103, 249–254.

Schlesselman, J.J. (1982) Basic methods of analysis. In Schlesselman, J.J. (ed.) *Case-control studies. Design, Conduct, Analysis*. Monographs in epidemiology and biostatistics. Oxford University Press. New York, Oxford.

Scotland, S.M., Said, B., Thomas, A. and Rowe, B. (1994) Ability of human sera to neutralise the activity of Verocytotoxins VT1, VT2 and variant forms of VTv. *FEMS Microbiology Letters* 115, 282–90.

Sharp, J.C.M., Coia, J.E., Curnow, J. and Reilly, W.J. (1994) *Escherichia coli* O157 infections in Scotland. *Journal of Medical Microbiology* 40, 3–9.

Shere, J.A., Bartlett, K.J., Kaspar, C.W. (1998) Longitudinal study of Escherichia coli O157: H7 dissemination on four dairy farms in Wisconsin (1998). *Applied and Environmental Microbiology* 64 (4), 1390–9.

Sherman, P., Cockerill, F., Soni, R. and Brunton, J. (1997) Outer membranes are competitive inhibitors of *Escherichia coli* O157:H7 adherence to epithelial cells. *Infection and Immunity* 59, 890–9.

Shukla, R., Slack, R., George, A., Cheasty, T., Rowe, B. and Scutter, J. (1995) *Escherichia coli* O157 infection associated with a farm visitor centre. *Communicable Disease Report* 5(R6), R86–R90.

Spika, J.S., Parsons, J.E., Nordenberg, D., Wells, J.G., Gunn, R.A. and Blake, P.A. (1986) Hemolytic uremic syndrome and diarrhea associated with *Escherichia coli* O157:H7 in a day care center. *The Journal of Paediatrics* 109, 287–91.

Steele, B.T., Murphy, N., Arbus, G.S. and Rance, C.P. (1982) An outbreak of HUS associated with ingestion of fresh apple juice. *The Journal of Pediatrics* 101, 963–5.

Stewart, P.J., Desormeaux, W. and Chene, J. (1983) Hemorrhagic colitis in a home for the aged – Ontario. *Canada Diseases Weekly Report* 9–8, 29–32.

Swerdlow, D.L., Woodruff, B.A., Brady, R.C., Griffin, P.M., Tippen, S., Donnell, H.D., Geldreich, E., Payne, B.J., Meyer, A., Wells, J.G., Greene, K.D., Bright, M., Bean, N.H. and Blake, P.A. (1992) A waterborne outbreak in Missouri of *Escherichia coli* O157:H7 associated with bloody diarrhoea and death. *Annals of Internal Medicine* 117, 812–19.

Synge, B.A. and Hopkins, G.F. (1992) Verotoxigenic *Escherichia coli* O157 in Scottish calves. [Letter] *The Veterinary Record* 130, 583.

Synge, B.A., Gunn, G.J., Ternent, H.E., Hopkins, G.F., Thomson-Carter, F., Foster, G., McKendrick, I. (2000) Preliminary Results from Epidemiological Studies in Cattle in Scotland. Proceedings from Zoonotic Infections in Livestock and the Risk to Public Health VTEC O157, Royal College of Physicians of Edinburgh (28th June).

Tildon, J., Young, A.M., McNamara, M.C. Symposium on new and emerging pathogens: *Escherichia coli* O157 – beyond the burger. In Abstracts of the 35th Interscience Conference on Antimicrobial Agents and Chemotherapy, San Francisco, California, September 17–20, 1995. Washington DC: American Society for Microbiology, 1995: 355.

Thomas, A., Chart, H., Cheasty, T., Smith, H.R., Frost, J.A. and Rowe, B. (1993) Vero cytotoxin-producing *Escherichia coli*, particularly serogroup O157, associated with human infections in the United Kingdom: 1989–1991. *Epidemiology and Infection* 110, 591–600.

Thomas, A., Cheasty, T., Frost, J.A., Chart, H., Smith, H.R. and Rowe, B. (1996) Vero cytotoxin-producing *Escherichia coli*, particularly serogroup O157, associated with human infections in England and Wales. *Epidemiology and Infection* 117, 1–10.

Thompson-Carter, F.M., Allison, L.J. and Pennington, T.H. (1996) *Escherichia coli* O157 in Scotland. *SCIEH Weekly Report* 30(13), 70.

Trevena, W.B., Hooper, R.S., Wray, C., Willshaw, G.A., Cheasty, T. and Domingue, G. (1996) Vero cytotoxin-producing *Escherichia coli* O157 associated with companion animals. *The Veterinary Record* 136, 400.

Trevena, W.B., Willshaw, G.A., Chesasty, T., Domingue, G., and Wray C. (1999) Transmission of Vero cytotoxin producing Escherichia coli O157 infection from farm animals to humans in Cornwall and West Devon. *Communicable Disease and Public Health* 2 (4), 263–8.

Tzipori, S., Karch, H., Wachsmuth, I.K., Robins-Browne, R.M., O'Brien, A.D., Lior, H., Cohen, M.L., Smithers, J. and Levine, M.M. (1987) Role of a 60-Megadalton plasmid and shiga-like toxins in the pathogenesis of infection caused by enterohemmorhagic *Escherichia coli* O157:H7 in gnotobiotic piglets. *Infection and Immunity* 55, 3117–25.

Tzipori, S., Gigson, R. and Montanaro, J. (1989) Nature and distribution of mucosal lesions associated with enteropathogenic and enterohemorrhagic *Escherichia coli* in piglets and the role of plasmid-mediated factors. *Infection and Immunity* 57, 1142–50.

Wadolkowski, E.A., Burris, J.A. and O'Brien, A.D. (1990) Mouse model for colonization and disease caused by enterohemmorhagic *Escherichia coli* O157:H7. *Infection and Immunity* 58, 2438–45.

Wall, P.G., McDonnell, R. J., Adak, G. K., Cheasty, T., Smith, H. R. and Rowe, B. (1996) General outbreaks of Vero cytotoxin producing *Escherichia coli* O157 in England and Wales from 1992 to 1994. *Communicable Disease Report Review* 6(2), R26–R33.

Waters, J.R., Sharp J.C.M. and Dev, V.J. (1994) Infection Caused by *Escherichia coli* O157:H7 in Alberta, Canada, and in Scotland: A Five-Year Review. *Clinical Infectious Diseases* 19, 834–43.

Wells, J.G., Davis, B.R., Wachsmuth, I.K., Riley, L.W., Remis, R.S., Sokolow, R. and Morris, G.K. (1983) Laboratory investigation of hemorrhagic colitis outbreaks associated with a rare *Escherichia coli* serotype. *Journal of Clinical Microbiology* 18(3), 512–20.

WHO (1996) Enterohaemorrhagic *Escherichia coli* infection. *Weekly Epidemiological Record* 35, 267–8.

Willshaw, G.A., Thirlwell, J., Jones, A.P., Parry, S., Salmon, R.L. and Hickey, M. (1994) Vero cytotoxin-producing *Escherichia coli* O157 in beefburger linked to an outbreak of diarrhoea, haemorrhagic colitis and haemolytic uraemic syndrome in Britain. *Letters in Applied Microbiology* 19, 304–7.

Wray, C., McLaren, I.M. and Carroll, P.J. (1993) Escherichia coli isolated from farm animals in England and Wales between 1986 and 1991. *Veterinary Record* 133 (18), 439–42.

Index

Note: Figures and Tables are indicated by *italic page numbers*

abattoir wastes
  spreading on agricultural land 102, 106, 111, 116, 122
  see also slaughterhouses
acid resistance (of VTEC O157) 79–80, 82, 135
adherence mechanism 5–6
Advisory Committee on the Microbiological Safety of Food (ACMSF) 28, 83–4
  Working Group on VTEC 76, 84
    recommendations 85, 90, 93–4, 98, 102, 126, 130–3, 141, 148
Agricultural Development Advisory Service (ADAS), Safe Sludge Matrix 102, 116, 122
agricultural land
  animal slurry spread on 87, 98, 99, 106
  recreational use 108, 118, 126
  sewage sludge spread on 102, 104, 116, 119, 122
analytical studies
  types *42*
  see also case control studies
animal manure
  contamination by
    of drinking water 104–10
    of fruit and vegetables 97–101
    of milk 95
    of raw meat/carcasses 89–93
    of recreational water 71, 110–14
  direct spread of VTEC O157 to humans via 69, 125–8
animal reservoirs 76–9, 82
animal transportation 91
antibiotics
  in animal husbandry 87
  VTEC O157 treated by 1, 12–13
antibody response 22–3
apple juice 62, 79, 80, 99
ascertainment bias 45
association (between exposure and disease) 43–4
  effect of bias 45–6
  effect of confounding variables 44–5
  meaning of term 152
  occurrence by chance 44
  power of study affecting 46–7
asymptomatic infection 25
attachment-and-effacement mechanism 6
attack rates 58

Bacillus cereus, pathogenic mechanism *6*
barbecues 131
bathing, outbreaks associated with 113, 124
bathing waters
  designated 112, 123
  microbiological quality 112, 121
Bathing Waters Directive 112, 121
beef products, outbreaks associated with 58–60
beefburgers
  outbreaks associated with 4–5, 9, 14, 48, 49, 58, 59–60, 70, 134
  recommendations on cooking 131–3, 135
  survey for VTEC 81
bias, in case control studies 45–6
bibliography 163

biphasic epidemic curve 5
bloody diarrhoea 12
bowel necrosis 11, 156
bowel perforation 4, 11, 156
British Pediatric Surveillance Unit (BPSU) 30, 31
burgers *see* beefburgers
butcher's shops
  licensing of 140, 141–2
  outbreaks associated with 65–6, 67, 144

Campylobacter spp, incidence of infection *3*, 39, 40
Canada
  case control studies 49
  incidence of VTEC O157 35, 37
  outbreaks 5, 58, 64
case control studies *42*, 152
  advantages 41–2
  analysis and interpretation of 43–7
  bias in 45–6
  compared with cohort studies *42*, 73–4
  control groups for 45–6
  disadvantages 42
  for sporadic VTEC O157 infections 47–9, 110, 113, 120, 124, 134
  threats to validity 45
case-searching 53, 72
case tracing 74
caterers, cooking instructions for beefburgers 132–3
cattle
  outbreaks associated with 48, 69
  as source of infection 76–9, 82, 97, 128
cell culture assay, vero cytotoxin detected by 20
Centers for Disease Control (CDC, USA)
  contact details 158
  laboratory investigations 4
  surveillance system 35
Central Public Health Laboratory (CPHL), diagnostic test(s) 21
cesspool waste 104
cheese
  outbreak associated with 61
  survey of 81
  unpasteurised 62, 93, 95
cheese curd, fresh, outbreak associated with 62
chemical treatments, in food production 100
children
  incidence of infections *32*, 35–6
    factors affecting 15
  secondary infection/transmission to/from 54, *55*, 70, 150
    advice on preventing 55–6
chlorinated water
  contamination of 109, 119
  washing with 100
Clostridium perfringens, pathogenic mechanism *6*
cluster sampling 74
cohort studies 38, *42*
  compared with case control studies *42*, 73–4
Communicable Disease Surveillance Centre (CDSC) 29, 53, 158
  contact details 158
companion animals, outbreaks associated with 50
confidence interval (CI) 43–4, 153
confounding 44, 153
  effects 44–5
consumer advice, on cooking beefburgers 131
consumer behaviour, changes in 14–15
consumer education 134, 143–4, 149
control groups 38–9, 45–6
Control of Pollution (Silage, Slurry and Agricultural Fuel Oil) Regulations 1991 106, 111
Control of Substances Hazardous to Health (COSHH) Regulations 1999 125, 146
controlled waters, meaning of term 106, 111, 115, 121
cooked meats
  cross-contamination of 69, 144
  outbreaks associated with 63, 65–6, 144
cooking instructions, for beefburgers 131–3
cows *see* cattle
cows' milk *see* milk
critical control points 83
  contamination of drinking water
    by animal manure 104–10
    by human sewage 114–20

contamination of fruit and vegetables
by animal manure 97–101
by human sewage 101–4
contamination of milk 93–7
contamination of raw meat 89–93
contamination of recreational water
by animal manure 110–14
by human sewage 120–4
cross-contamination of ready-to-eat foods 69, 138–45
heat treatment (thorough cooking) 10, 48, 52, 69, 129–37
on-farm control 85–8
person-to-person transmission 145–51
zoonotic spread 125–8
cross-contamination 61, 67, 68–9, 95
prevention of 69, 138–45
cross-sectional studies *42*
Cryptosporidium spp, incidence of food poisoning due to *3*
CR-SMAC method 18–19
advantages *19*
disadvantages 18, *19*
cut-off value 44

dairy establishments (milk processing) 95
Dairy Hygiene Inspectorate 93, 158
dairy products
hygienic production of 93–7
outbreaks associated with 60–2
Dairy Products (Hygiene) Regulations 1995 93, 95
data, analysis and presentation of 73
deer, as source of infection 87, 128
diarrhoea-causing pathogens 1, 2–4
dichotomous variable 43
diffusely adherent *E.coli* (DAEC) 2, 3
DNA probe test 20–1
drinking water
contamination of
by animal manure 104–10
by human sewage 114–20
enforcement arrangements on quality 105–7, 115–16
private supplies 106–7, 108, 109, 116, 117, 118, 119
sources 109, 118
treatment of 109, 118
Drinking Water Inspectorate (DWI) 105, 115, 159

ecological studies *42*
educational programmes 134, 142, 143–4, 148, 149
elderly people, infection of *12*, 15, 65, 144
England
incidence of VTEC O157 *31*, *32*, *33*, 37
outbreaks 60–1, 67, 69, 113, 124
enrichment technique (in isolation of VTEC O157) 18–19
applications 79, 113, 124
Enternet surveillance system 30
enteroaggregative *E.coli* (EaggEC) *2*, 3
enterohaemorrhagic *E.coli* (EHEC) 3
WHO consultation on 85, 91–2, 94, 99, 103–4, 108, 113, 118, 123, 126–7, 134, 143–4, 149
see also VTEC O157
enteroinvasive *E.coli* (EIEC) *2*, 3
enteropathogenic *E.coli* (EPEC) 2
enterotoxigenic *E.coli* (ETEC) 2, 3
Environment Agency
contact details 159
enforcement by 98, 102, 105, 106, 111, 115, 116, 121, 122
enzyme linked immunosorbent assay (ELISA) 153
advantages and disadvantages 20
verocytotoxin detected by 20, 23
epidemic curves 5, 73, 153
epidemiological studies 4, 5, 25, 41, 73–4, 153
exposure measurement methods in 46
in waterborne outbreaks 109, 119–20
see also case control studies; cohort studies
Escherichia coli (*E.coli*) 1
disease-causing types 2–4
see also VTEC O157
European Directives 105, 112, 115, 121
excretion, period of 23–5
exposure measurement bias 46

faecal–oral transmission 54, 145–51
faecal specimens, examination of 17–19

faeces *see* human sewage
farm animals
  and control of VTEC O157 85–8, 125–8
  infections associated with 48, 50, 69
  as reservoir of infection 76–9, 87
farm visitor centres 69, 125
farm waste management plans 98, 106, 111
farm workers, Pennington Group recommendations 85, 90, 94, 98, 102, 107, 112–13, 117, 123, 126
fatality rates 11, *12*, 66
food handlers
  infected, transmission via 145–51
  training of 133, 134, 142, 143, 149
food hygiene inspections 130, 139–40, 147–8
food manufacturers, cooking instructions for beefburgers 131
food outlets/premises
  licensing of 140, 141–2
  outbreaks associated with 63–7, 144
  PHLS survey 81
  requirements on design 139, 142
food poisoning
  compared with VTEC O157 outbreaks 74
  definition 28
  incidence *3*
  VTEC O157 as 17, 148
food production, changes in 14
food reservoirs 79–81, 82
Food Safety Act 1990 129, 138, 146
  Codes of Practice 130, 139, 141, 147
Food Safety (General Food Hygiene) Regulations 1995 129, 132, 138, 141, 147
Food Standards Agency 83
  contact details 159–60
  Joint Task Force on *E.coli* O157 (in Scotland) 84
  Meat Hygiene Enforcement Report 89
foodborne outbreaks 4–5, 9, 14, 48, 49, 58–67, 68–9, 70, 134
  notification of 28
forage crops 102, 116, 122
Fresh Meat (Hygiene and Inspection) Regulations 1995 89, 90
fruit
  contamination of
    by animal manure 97–101
    by human sewage 101–4

gastrointestinal infections, international surveillance system for 30
gastrointestinal tract stricture 11, 154
general practitioners (GPs) 28–9, 37–8
glossary 152–7
goats
  milk from 95
  outbreaks associated with 69
  as source of infection 87, 128
ground beef
  infection associated with 49, 58
  VTEC in 80

haemolytic anaemia 10, 154
haemolytic uraemic syndrome (HUS) *2*, 3, 4, 5, 7, 8, 10–11
  reports of outbreaks 13
  surveillance system for childhood HUS 30–1
haemorrhagic colitis 3, 4, 5, 10, 29, 154
  first reported outbreak 4, 13, *59*
  as reportable disease in Canada 37
  seen by GP 29
hamburgers *see* beefburgers
Hazard Analysis and Critical Control Points (HACCP) system 144–5, 154
  in abattoirs/slaughterhouses 90, 91
  in food industry and food premises 93, 98, 133, 134, 140, 141, 142, 143, 144–5, 148–9
  principles 145
  see also critical control points
Health Mark 89–90
health risk condition 130, 139
Health and Safety Executive 125, 146, 160–1
Health and Safety at Work etc. Act 1974 123, 125, 146
heat treatment
  as critical control point 10, 48, 52, 69, 129–37
  VTEC O157 affected by 80, 95, 133
hepatitis 11
horses, as source of infection 87, 128
hospitals, outbreaks in 70, 150
hot dogs, outbreaks associated with 49
household spread of infection 54, *55*

human sewage
  contamination by
    of drinking water 114–20
    of fruit and vegetables 101–4
    of recreational water 120–4
Hygiene Assessment System (HAS) 89
hygiene procedures
  in domestic households 55–6
  at farm visitor centres 69, 125
  in slaughterhouses 89, 90
hypothesis testing 72–3

immunoglobulin response 22–3
immunomagnetic separation (IMS) technique 19, 154
  advantages and disadvantages *20*
  applications 79, 81, 113, 124
incidence of VTEC O157 1, *3*, 31–40
  in Canada and USA 33, 35, 37
  in Netherlands 39–40
  seasonal variation 31, *32*, 37
  in UK 31–3, *34*, 35–7, 39
incubation period 8–9, 154
infectious dose 9, 154
Infectious Intestinal Disease (IID) study 37–40
institutional establishments, outbreaks in *12*, 54, 70, 150
international surveillance network 30
international travel, effects 15
interviewer bias 46
interviews 72
Ireland, outbreak in 60
irradiation of food 99, 100, 103
irrigation water 99, 103

Japan, outbreak(s) in 63, *64*, 99
jerky, outbreaks associated with 135

kidney failure 1, 10

laboratory-acquired infections 9, 150, 151
Laboratory Centre for Disease Control (Canada) 13, 35
laboratory reports 29
lipopolysaccharide (LPS) 22, 155
livestock industry changes 13

Management of Health and Safety at Work Regulations 1992 123, 125
mayonnaise 80
Meat Hygiene Service 89
  Pennington Group recommendations 90, 91
Meat Products (Hygiene) Regulations 1994 66, 89, 138, 141, 144
microbiological investigations 17–19
  of outbreaks 69
  of sporadic infections 49–50
milk
  contamination of 93–7
  enforcement arrangements 93
  outbreaks associated with 60–2
  see also pasteurised . . .; unpasteurised milk
minced beef 80, 131
  see also beefburgers; ground beef
misclassification, bias caused by 46
multivariate analysis 45, 155

nappies, of infected children 56
Natural Mineral Water Regulations 1985 106, 116
Netherlands, gastroenteritis study 39–40
non-bloody diarrhoea 11
non-invasive diagnostic tests 23, 25
null hypothesis 44

odds ratio 43, 47, 155
on-farm control 85–8
outbreak reports 29
outbreaks 58–75
  causes 58–71, 75
  detection of 36, 53
  identification and confirmation of 71–3
  investigation of 58
    practical approach 71–4
  and sporadic cases 53
  VTEC O157 compared with food poisoning outbreaks 74

pancreatitis 11
pasteurisation, effect on VTEC O157 61, 81
pasteurised milk, cross-contamination of 61, 68
pasture
  guidance on recreational use 108, 118, 126
  sludge/slurry spreading on 102, 106, 116

Pennington Group 84
recommendations 85, 90–1, 94, 98, 102, 107, 112–13, 117, 123, 126, 133, 141–2, 148–9
Pennington Report 84
implementation of key recommendations 130, 140, 144
period of excretion 23–5
person-to-person transmission 5, 9, 24, 54, 65, 70
groups that pose special risk 54, 148, *150*
prevention of 54–6, 145–51
see also secondary transmission
phenotypic typing 21
pigs, as source of infection 76, 77, 82, 87, 128
plasmid analysis 22
point-source outbreaks 73, 155
polymerase chain reaction (PCR) method 21, 155
population-based surveillance 27–8, 155
data sources 28–30
early programmes 5
objectives 27
population at risk 72
power of study 47
private water supplies 106–7, 108, 109, 116, 117, 118, 119
disinfection of 109, 119
probability sampling procedures 46
production holdings (milk production) 95
Public Health Act 1936 122–3
Public Health (Control of Disease) Act (1984) 17, 148
Public Health Laboratory Service (PHLS)
co-operation with SCIEH 108, 118
contact details 161
Laboratory of Enteric Pathogens (LEP) 21, 23, 29, 36, 53
standard method 18, 79, 81
surveillance system(s) 29, 30, 31
pulsed field gel electrophoresis (PFGE) 22
applications 65, 67, 109

random sampling 74
raw meat
consumption of 129–37
contamination of 89–93
raw milk *see* unpasteurised milk
ready-to-eat foods, cross-contamination of 69, 138–45
ready-to-eat fruit and vegetables 63, 98, 99, 102, 116
recall bias 46
recreational use of agricultural land 108, 118, 126
recreational water
contamination of
by animal manure 71, 110–14
by human sewage 70–1, 120–4
enforcement arrangements on quality 111–12, 121–3
renal failure 1, 10
reportable diseases 17, 148
reservoirs of VTEC O157 76–82
residential care homes *12*, 150
restaurants, outbreaks associated with 63–4
restriction fragment length polymorphism (RFLP) 22
risk assessments, of food handling procedures 52–3
risk communication 134, 144, 149

Safe Sludge Matrix 102, 116, 122
salad crops 63, 98, 99, 102, 116
salami, dry-cured, outbreaks associated with 9, 62–3, 135
Salmonella spp.
incidence of infection *3*, 39, 40
pathogenic mechanism *6*
sampling techniques 52, 53, 74
Scotland
animal reservoirs in 78
incidence of VTEC O157 32–3, 37
Joint Task Force on *E.coli* O157 84
outbreaks 61, 63–4, 65–6, 68, 144
Scottish Centre for Infection and Environmental Health (SCIEH) 29, 30, 31, 161
co-operation with PHLS 108, 118
Scottish *E.coli* O157 Reference Laboratory 23, 30, 31, 162
Scottish Task Force on *E.coli* O157 84
evidence presented to 85
seasonal variation (in incidence of VTEC O157) 31, *32*, 37

secondary transmission 5, 70
among household members 54, *55*
prevention of 54
advice on 54–6
selection bias 45–6
selective/screening media 18
septic tank sludge, spreading on agricultural land 102, 104, 116, 119, 122
serology 22–3
sewage sludge
spreading on agricultural land 102, 116, 122
see also human sewage
sheep
milk from 95
as source of infection 77, 79, 82, 87, 128
Shigella spp.
contamination of water by 70
pathogenic mechanism *6*, 7
slaughterhouses
enforcement arrangements 89–90
hygiene in 89–90, 91–2
Sludge (Use in Agriculture) Regulations 1989 102, 116, 122
slurry
spreading on agricultural land 87, 98, 99, 106
storage facilities 98, 106, 111
see also animal manure
social behaviour, changes in 15
sorbitol fermentation 17
Sorbitol MacConkey Agar (SMA) 18, 109, 119
cefixime/rhamnose modified (CR-SMA) 18
sporadic infections 41–57
in Canada compared with USA 35
case control studies for 47–9
finding sources of 41–3
gathering information on potential sources 51–3
investigation of 50–6
advice to be given 54–6
aims 50–6
checklist for 57
microbiological investigation of 49–50
need for urgent investigation 50
as part of outbreaks 53
sprouting seeds
irrigation of 99, 103, 104
outbreaks associated with 63, 99
Staphylococcus aureus, pathogenic mechanism *6*
statutory notifications 28–9
strain discrimination 21–2
stratified sampling 74
surveillance 27–8
data sources 28–30
international network 30
laboratory reports 29
outbreak reports 29
statutory notifications 28–9
early programmes 5
information flow *28*
swimming pools
contamination of 70, 71, 124
monitoring of water quality 123

Task Force on *E.coli* O157 (in Scotland) 84
evidence presented to 85, 91, 98, 103, 107–8, 113, 117–18, 123, 126, 133, 143, 149
thrombotic thrombocytopenic purpura 11, 156
toilets, cross-infection via 55
training of food handlers 133, 134, 143, 149
Type I error (false-positive) 47
Type II error (false-negative) 47

UK
incidence of VTEC O157 infections 31–3, *34*, 35–7, 39
outbreaks 60–2, 63–4, 65–6, 68, 69, 113, 124
ultraviolet disinfection units, for drinking water 109, 119
undercooked meat, consumption of 129–37
univariate analysis 45
unpasteurised milk
outbreaks associated with 60–1
products from 62, 94
sale and consumption of 93, 95
USA
incidence of VTEC O157 infections 33, 35, 37
outbreaks 4–5, 58–60, 62–3, 68, 70, 134

vegetables
  contamination of
    by animal manure 97–101
    by human sewage 101–4
  outbreaks associated with 62, 63, 99
  restrictions on sewage sludge
      spreading 102, 116
  VTEC O157 on 80
vero cytotoxin-producing bacteria 1, 3,
      7, 7–8, 19
  see also VTEC O157
vero cytotoxins 1, 3, 7, 19
  identification of 19–21
  physiological action 1, 3, 7, 19
  types 7, 19, 22
Veterinary Laboratory Agency 52, 162
Vibrio spp., pathogenic mechanism *6*
VTEC O157
  acid resistance 79–80, 82, 135
  and antibiotic treatments 1, 12–13
  characteristics 17, 18
  control of 83–151
    in cooked foods 129–37
    in drinking water 104–10, 114–20
    on farms 85–8
    in fruit and vegetables 97–104
    in milk 93–7
    and person-to-person
      transmission 138–45
    in raw meat 89–93
    in ready-to-eat foods 69, 138–45
    in recreational water 110–14,
      120–4
    and zoonotic spread 125–8
  diagnosis of 17–26
    future developments 25
  disease/clinical symptoms 10–13
  effect of heat treatment 80, 95
  emergence of 4–5, 13
    factors influencing 13–15
  first recorded outbreaks 4–5, 58, 134
  heat resistance 80, 95, 133
  incidence of infection 1, *3*, 31–7, 39
  incubation period 8–9
  infectious dose 9, 15
  isolation of 17–19
  low-temperature resistance 80–1
  meaning of nomenclature 2
  mechanism of causing illness 5–8
  outbreaks 58–75
    causes 58–71, 75
    investigation of 71–4
  pathogenic mechanism *6*
  period of excretion 23–5
  as reportable disease 17, 148
  reservoirs 76–82
  severity of illness, outbreak reports
      12
  spread of
    outbreaks 58–75
    sporadic cases 41–57
  survivability in water 109, 113, 119,
      123
  toxin production by 7–8
    identification of verotoxin 19–21
  transmission pathways *86*
  virulence attributes 5–8
    practical implications 9–10

Wales
  incidence of VTEC O157 *32*, 33, *34*,
      35–6, 37
  outbreaks 33, *34*, 67
Waste Management Licensing
      Regulations 1994 102, 104, 106,
      111, 116, 122
water *see* drinking . . .; irrigation . . .;
      recreational water
Water Code (slurry storage facilities)
      106, 111
Water Industry Act 1991 105, 115
Water Resources Act 1991 105, 111,
      115, 121
Water Supply (Water Quality)
      Regulations 1989 105, 115
waterborne outbreaks 9, 68, 109, 119
well-water 109, 119
  contamination of 108, 118
wholesale catering suppliers, cooking
      instructions for beefburgers 132
World Health Organization (WHO) 28
  consultation on EHEC 85, 91–2, 94,
      99, 103–4, 108, 113, 118, 123,
      126–7, 134, 143–4, 149
  on consumer education 134, 143–4,
      149
  contact details 162
  on training of food handlers 134,
      143, 149

yoghurt 61–2, 95

zoonotic spread 69
  prevention of 125–8